D0850582

Evolution and Consciousness

HUMAN SYSTEMS IN TRANSITION

Evolution and Consciousness

HUMAN SYSTEMS IN TRANSITION

Edited by

Erich Jantsch and **Conrad H. Waddington**

1976

Addison-Wesley Publishing Company

Advanced Book Program

Reading, Massachusetts

London • Amsterdam • Don Mills, Ontario • Sydney • Tokyo

The cover of this volume shows the nucleation of a new macron, photographed by Ralph Abraham with the macroscope of the University of California, Santa Cruz. (See also Chapter 6, p. 144.)

Library of Congress Cataloging in Publication Data

Main entry under title:

Evolution and consciousness.

1. Evolution — Addresses, essays, lectures. 2. Man — Addresses, essays, lectures. 3. Consciousness — Addresses, essays, lectures. 4. System theory — Addresses, essays, lectures. 5. Social systems — Addresses, essays, lectures. I. Jantsch, Erich. II. Waddington, Conrad Hal, 1905-
B818.E83 128 76-25102
ISBN 0-201-03438-7
ISBN 0-201-03439-5 pbk.

Manufactured in the United States of America

ABCDEFGHIJ-AL-79876

The world was spinning like the most delicately tinted of bubbles, all light. It was the mind of humanity that I saw, but this was not at all to be separated from the animal mind which married and fused with it everywhere. Nor was it a question of higher or lower . . . . I watched a pulsing swirl of all being, continually changing, moving, dancing, a controlled impelled dance, held within its limits by its nature, and part of this necessity was the locking together of the inner pattern in light with the other world of stone, leaf, flesh and ordinary light . . . .

And on this map or plan that showed how myriads of ridiculously self-important identities were reduced to a few, was another, different, but, in some places, matching pattern, of a stronger, rarer light (or sound) that varied and pulsed and changed like the rest but connected direct, made a link and a bridge, a feeding channel, between the outer (or inner, according to how one looked at it) web of thought or feeling, the pulsating bubble of subtle surrounding colour, and the solid earthy watery globe of Man. Not only a link or a bridge merely, since this strand of humanity was open like so many vessels open to the rain, but part of the shimmering web of fluid joyful being, which was why the scurrying, hurrying, scrabbling, fighting, restless, hating, wanting little patches of humanity, the crusts of lichen or fungi growing here and there on the globe, the sea's children, were, in spite of their distance from the outer shimmering web, nevertheless linked with it always, since at every moment the glittering tension of singing light flooded into them, into the earthy globe, beating on its own delicious pulse of joy and creation.

Doris Lessing, *Briefing for a Descent into Hell*

Contents

Acknowledgments

Chapter 4 is partly based on extracts from C. S. Holling, "Resilience and Stability of Ecological Systems," *Annual Review of Ecology and Systematics*, Vol. 4 (1973), pp. 1-23. Chapter 5 is based on a presentation by Ilya Prigogine to the Collège de France, Paris, December 1974. Chapter 11 by O. W. Markley is based on a study of the societal consequences of changing images of man which was supported by the Charles F. Kettering Foundation. Chapter 12 is partly based on a presentation by Erich Jantsch to the Third International Conference on the Unity of the Sciences, "Science and Absolute Values, " London, November 1974, which is published under the title "The Quest for Absolute Values" in the conference proceedings (International Cultural Foundation, Tarrytown, N. Y., Tokyo, and London) as well as in *Futures*, Vol. 7 (1975), 463-474.

The quotation from Doris Lessing's *Briefing for a Descent into Hell* which forms the motto of this volume, is used by permission of Alfred A. Knopf, Inc., New York.

The C. G. Jung quotation which serves as motto for Part III is from C. G. Jung, "Ein Brief zur Frage der Synchronizität," *Zeitschrift für Parapsychologie und Grenzgebiete der Psychologie*, Vol. 5, No. 1 (1961), pp. 4f.; it is quoted in English in Marie-Louise von Franz, *Number and Time* (English translation by A. Dykes), Evanston, Ill.: Northwestern University Press (1974), p.8.

The cover photograph, "Nucleation of a Macron," has been made by Ralph Abraham with the macroscope of the University of California, Santa Cruz.

Contributors

Ralph Abraham (pp. 134–149)	is Professor at the Division of Natural Sciences, University of California, Santa Cruz.
Peter M. Allen (pp. 127–130)	is with the Faculté des Sciences, Université Libre de Bruxelles, Brussels.
C. S. Holling (pp. 73–92)	is Professor at the Institute of Resource Ecology, University of British Columbia, Vancouver.
Erich Jantsch (pp. 37–72, 230–242)	is with the Center for Research in Management Science, University of California, Berkeley; he is also a frequent visiting professor at various American and European universities.
O. W. Markley (pp. 214–229)	is with the Center for the Study of Social Policy, Stanford Research Institute, Menlo Park, California.
Milton Marney (pp. 185–197)	is with the Program of Policy Studies in Science and Technology, George Washington University, Washington, D. C.
Magoroh Maruyama (pp. 198–213)	is Professor at the Systems Science Ph.D. Program, Portland State University, Portland, Oregon.
Walter Pankow (pp. 16–36)	is with the Nationalfonds-Projekt "Wachstum-Umwelt," Eidgenössische Technische Hochschule (ETH), Zurich; while writing his contribution to this volume, he was also with the Institute of Interdisciplinary Biology, University of Essen.
Norbert A. Pierre (pp. 150-165)	is with the Graduate School of Business, Columbia University, New York.

Ilya Prigogine *(pp. 93–126, 130–133)* is Professor at the Faculté des Sciences, Université Libre de Bruxelles, Brussels, and Professor of Physics and Director of the Center for Statistical Mechanics and Thermodynamics, The University of Texas, Austin.

Paul F. Schmidt *(pp. 185–197)* is Professor at the Department of Philosophy, University of New Mexico, Albuquerque.

Alastair M. Taylor *(pp. 169–184)* is Professor at the Department of Geography, Queen's University, Kingston, Ontario.

Conrad H. Waddington(†) *(pp. 11–15, 243-249)* was Professor at the Institute of Animal Genetics and Director of the School of the Man-Made Future, University of Edinburgh, Scotland.

Milan Zeleny *(pp. 150–165)* is Professor at the Graduate School of Business, Columbia University, New York.

†died September 26, 1975

Introduction and Summary

It is the business of the future to be dangerous. . . .
The major advances in civilization are processes that
all but wreck the societies in which they occur. . . .

Alfred North Whitehead, *Adventures in Ideas*

In our time, the clash between the two dominant myths in the dualistic world view of Western man becomes accentuated again in a most dramatic way: *Prometheus*, who stole the fire from heaven and set out to establish mankind as a creed of counter-gods, is still the hero of the official Western ideology of progress and dominance through technology; whereas *Sisyphus*, who struggles to roll his fate on to a platform of eternal rest, but has to recommence over and over again, turns out to be the involuntary, sad hero of all those frustrated attempts to create a lasting structure somewhere, anywhere in the human world: from Marxism with its promise of classless society over Keynes (who speculated that postindustrial society will get along without economy) and the quickly vanishing post-World War II confidence in Western science as the monolithic religion for a coming world culture, all the way to America's missionary zeal in "bringing democracy to the world," governments' avowed purpose to preserve the institutions of society, and—coming to prominence in our day—the ideology of enforced structural stabilization behind world models cast in rigid behavioral and equilibrium terms. The futility of such attempts to stop the world is always blamed on external forces—which tends to alienate us further from our own world, which is becoming increasingly elusive and callous to *our* idea of imposed control.

But Prometheus and Sisyphus are not the only myths to choose from. In fact, they are bound to Western culture. There is another myth which has illuminated human life throughout the millennia and across many cultures, so that Aldous Huxley has rightly called it "the perennial philosophy." This is the myth holding that human life is sharing integrally in a greater order of process, that it is an aspect as well as an

ISBN 0-201-03438-7/0-201-03439-5pbk.

Erich Jantsch and Conrad H. Waddington (eds.), *Evolution and Consciousness: Human Systems in Transition.*

agent of universal evolution. The fear of losing static security in a given structure and of being swept along by an unpredictable stream thus becomes transformed into *hope*—the hope associated with *life*, with the dynamic notions of continuity and transformability, of being embedded in a purpose and meaning transcending ourselves and the lives of our transitory systems, the hope inherent in the nondualistic experience of *being* the stream.

This volume attempts to bring together some of the emergent concepts and approaches which seem capable of developing into elements for a scientific foundation to this kind of hope. The evolutionary paradigm is still almost totally neglected by a social science which finds its purpose in reducing the human world to the equilibrium perfection, structural unambiguity and permanence, hierarchical control, and predictability of machine-like structures. In contrast, the contributions to this book try to develop a new understanding of an evolving world of human systems which are characterized by the same aspects of imperfection, nonequilibrium, and nonpredictability, of differentiation and symbiotic pluralism, which seem to govern life in all its manifestations. They argue that the human world, analogous to physical and biological evolution, incorporates a basic principle of *self-transcendence*, of venturing out by changing its own physical, social, and cultural structures—above all, by changing its own consciousness. What is real in this exciting drama of continuous self-renewal and self-expression are self-bounding *processes* rather than the elusive and transitory structures which arise from their interaction in ever-changing forms and complexities.

In this perspective, any attempt to stabilize structures by imposing boundaries—the *ultima ratio* of the machine syndrome—cannot but become counterproductive in two ways. First, a stabilizing control imposed *upon* the world and interfering with processes of self-transformation and self-organization is inevitably bound to end up with dictatorship; equilibrium is synonymous with social and cultural, and ultimately also physical, death. Second, if the human world cannot be put to death in such a way—and I believe that no religious, political, or scientific zealot will ever totally triumph over life—enforced stabilization and equilibration will amount to but a temporary halting of the processes of life, which will ultimately break through with vastly increased explosive and disruptive power, just as a boiler will burst if it is expected to contain in the same structure water transforming into steam. The false paradigms and expectations of stabilization are becoming a serious threat as our century enters its last quarter. Therefore, this book also tries to convey a certain sense of urgency, an appeal to trust in the evolving gestalt of life rather than in lifeless form.

ISBN 0-201-03438-7/0-201-03439-5pbk.

The book is divided into three parts:

Part I deals with the evolutionary paradigm in an emergent perspective. *Chapter 1* by *Waddington* summarizes the recent thrust of evolutionary theory in the subhuman domain. The increasing flexibility of the phenotype (the individual) and selection through mutual correlation between a whole population of organisms and its environment appear as core principles of open-ended sociogenetic processes.

Chapter 2 by *Pankow* discusses the basic characteristics of living systems and the possibility of their representation through science. It introduces the fundamental difference between a living gestalt system and a formal system by which it is represented. The openness of gestalt systems goes beyond the formal openness to flows of energy, matter, and information; it is self-transcendence, the capability of a system to represent itself. Only gestalt can recognize gestalt. Therefore, building up increasingly sophisticated ways of representation through formal languages (science) can never grasp a gestalt system in its entirety; only natural language, itself a gestalt system, can do this.

In *Chapter 3*, *Jantsch* tries to link the concepts developed in this volume by discussing evolution as self-realization through self-transcendence. Dynamic process thinking—in contrast to equilibrium-oriented structural thinking—leads to a generalized concept of evolutionary experimentation, the venturing out and subsequent vindication through open, intermeshing learning processes which result in the mutual correlation of species with the physical environment, and in the human world also of man with his social and cultural environments. A basic principle in human evolution is again flexibility in adaptation and action, now generally of the "system phenotype" from individuals to social and cultural (noetic and consciousness) systems. A few general conclusions are drawn for human design: structural flexibility, amelioration of natural processes, and quasi-continuous "dynamic balance," that is, the design of transitions between temporary stable regimes which may be expected to be less disruptive than unattended switches.

Part II deals with formal approaches to evolving systems which are based on process thinking and therefore, in contrast to structure-oriented mechanical and behavioral approaches, are capable of representing to some degree that basic characteristic of life which expresses itself as "continuity in change." *Chapter 4* by *Holling* concludes from empirical studies of natural ecological systems that their dynamics cannot be explained by an equilibrium-centered view. Resilience, or an ecosystem's capability to persist through changes, seems to be generally linked to high fluctuations (or low local stability); attempts to stabilize certain ecological systems through human management have resulted in practically ruining them. In theoretical terms, the concept of multiple

ISBN 0-201-03438-7/0-201-03439-5pbk.

stable regions in phase space, or domains of attraction, leads to a general nonequilibrium or boundary-oriented view of ecosystems. Again, what evolution seems to maximize is not efficiency or productivity, but flexibility to persist. In the examples from natural ecology, persistence refers to a given dynamic regime. It is conjectured that high resilience through high fluctuations also enhances the persistence of a system undergoing qualitative change by switching regimes. If this is so, the application of these insights to the human world (which currently is geared to the maximization of efficiency and productivity) suggests a strategy which ensures that human systems keep as close as possible to a stability boundary, emphasizing nonequilibrium, variation, and fluctuation—a total reversal of conventional wisdom.

Nonequilibrium emerges as a principle inherent in life also in *Chapter 5* by *Prigogine.* Discussing systems of interactive populations—systems with a history—in a sweeping view from physical-chemical systems through animal aggregations and societies to natural ecosystems and further to human sociocultural systems, the concept of "order through fluctuation" emerges as a basic nonequilibrium ordering principle governing dynamic aspects of evolving systems at many levels. This newly recognized Prigogine principle forms the core of the most comprehensive approach to evolving systems so far developed. It views the role of fluctuations, amplified through systems, as both defining (enhancing) and transforming space-time structures; in fact, dissipative structures may themselves be described as giant fluctuations. "Order through fluctuation" is discussed from two points of view. A macroscopic systems view focuses on partially open, sufficiently nonequilibrium "dissipative structures" which, in contrast to equilibrium structures (the other type of structure occurring in nature), cannot exist independently and are maintained in exchange with their environment; the resulting complementarity between function and structure of such a system is also emphasized in many of the other chapters. Another view focuses on a stochastic formulation in order to study the consequences of a specific (typically random) fluctuation and the conditions for its penetration or its being damped by the environment. Human systems, due to their richly coupled subsystems, may generally be expected to live in a state of metastability, absorbing relatively big fluctuations without being forced to another temporary stable regime—thus living, in Holling's terms, close to a stability boundary where resilience is high.

Chapter 6 by *Abraham* discusses macrodynamics and morphogenesis in the light of still other recent theoretical and experimental developments, namely, the theory of catastrophes (Thom) and the observations made with the macroscope. An example of the latter is used for the cover of this volume. Simple physical, chemical, and electrical macrons—stable regimes of dynamic systems—and complex systems of

ISBN 0-201-03438-7/0-201-03439-5pbk.

macrons are discussed in terms of their transitions (catastropnes) from one stable regime to another. But since structural descriptions ("attractors") are used for macrons, the role of fluctuations and their amplification in self-organization is not accounted for by this theoretical approach. Applications of the results of experimental macroscopy to a wide variety of fields, from cosmology and geology through biology and neurophysiology to noology (mental and consciousness systems) seem promising.

In *Chapter 7*, finally, *Zeleny* and *Pierre* discuss recently developed approaches to the simulation of interactive natural processes characterizing self-renewing systems (such as cells) with a view to their potential application to human systems management. In such a perspective, a manager would act as a catalyst, furthering self-organizing processes, rather than as a designer, controller, or analyst of the system.

Part III deals with selected aspects of sociocultural evolution in the human domain. *Chapter 8* by *Taylor* discusses process and structure in sociocultural systems with the help of cybernetic models of a single system, of two geopolitical systems in interaction, and of the successive stages of emergent geopolitical systems with an impending switch from nation-state to planetary scope. The initiation of "order through fluctuation" is discussed, for historical societies as well as for our present situation, in terms of positive feedback loops provided by these cybernetic models. As a useful concept for the theory of geopolitical evolution, "space as plenum," or space as an ordering constituent of macroscopic field-forces, is introduced.

In *Chapter 9*, *Marney* and *Schmidt* present the cognitive modality, or method of scientific inquiry, as an evolving system moving from the axiomatic through the empirical and the constructural "stable state" to the contemporary modification (normative method), outlining at the same time the "fluctuations" which forced a qualitative change to a new regime. As emergent features of scientific advance, or principles of noetic evolution, may be recognized the maximization of cognitive freedom and scope and the optimization of cognitive control. The quest for certainty is gradually relinquished and flexibility in terms of long-term viability moves into focus.

Chapter 10 by *Maruyama* discusses from an anthropologist's point of view the development of heterogeneity, or pluralism, in sociocultural systems which is being ordered by the nonequilibrium, complementary principle of symbiosis. In this context, a dialogue between epistemologies and paradigms becomes imperative. An emerging mutual-causal logic, new to the Western world, may be expected to mark a transition from a chain of temporary stable regimes toward a metaregime of perpetually transforming patterns—a transition between types of transition and thus a metatransition.

ISBN 0-201-03438-7/0-201-03439-5pbk.

Chapter 11 by *Markley* traces the transformation of human consciousness by briefly discussing five typical examples: cultural revitalization, revolutions in science (within Marney's and Schmidt's "constructural prototype"), heroic mythology, psychotherapy, and general creativity. The basic transformation cycle has (so far) always been the same, characterized by a nonincremental switch to a new regime which, in these cases, emerged as a new guiding image. For our present situation, features of a coming knowledge paradigm and a coming image of man (cultural paradigm) are briefly outlined in terms of their necessity as well as their actual emergence.

Chapter 12 by *Jantsch* finally discusses the evolution of images of man (or, more precisely, man-in-universe) in connection with the processes of integration (*re-ligio*) and differentiation (history) in three "waves": grounding, socialization, and individuation. Images of man tend to spring up long before history comes to live them out, thus illuminating and guiding the mankind process over thousands of years ahead. The basic image of individuation which was anticipated an aeon ago will have to be lived out in the impending Age of Aquarius.

From the very rich discussions offered in these twelve chapters, some general characteristics seem to emerge which will become important for a deeper understanding of the nature of evolving systems (or systems with a history, described by variables with a high degree of interdependence) in general, and human systems in particular. These characteristics include the following:

1. A state of sufficient *nonequilibrium* is maintained within the system and in its relations with the environment.
2. *Functions* (the relations with the environment) and *structure* determine each other; they are complementary.
3. *Deterministic* and *stochastic* (random) features are interdependent; chance and necessity become complementary in a process view.
4. There exist *multiple stable regions*, or dynamic regimes, for the system; in switching between them, the system has the capability of undergoing *qualitative change.*
5. Evolution implies an ordered *succession* of such *transitions;* autocatalysis at many levels seems to be a principal "driving mechanism."
6. *Resilience,* or persistence (metastability), is high near the boundaries of a stable regime (i.e., near the maximum sustainable nonequilibrium), where fluctuations abound and stability (*qua* return to an equilibrium state) is low; inversely, high stability generally implies low resilience—a system geared to short-term efficiency and productivity.
7. The thrust of evolution seems to further *flexibility* of the "system phenotype" (the individual system) at all levels; this implies that long-term viability through the development of a capability to deal with the unexpected is favored over short-term efficiency and productivity.

ISBN 0-201-03438-7/0-201-03439-5pbk.

8. By virtue of this flexibility, the evolutionary processes work through *evolutionary experimentation* at many levels of an open learning hierarchy—testing directions, not places, and finding confirmation *a posteriori* through *vindication* (and not *a priori* through certainty and prediction).
9. The result of evolution is a *progressive correlation between genotypes and physical, social, and cultural environment,* with adaptation of the specific genotype to a specific level of environment in the first phase, and adaptation of the environment to the genotype in the second phase. This progressive correlation is characterized by an increasing emphasis on epigenetic group processes, mainly intersystemic exchange, which in the human world come to dominate over linear genetic processes.
10. A basic principle of this correlation is *symbiotization of heterogeneity.*
11. High resilience through high fluctuation may be assumed to imply an enhanced capability not just for persistence within a particular regime, but also for *long-term viability through transformability.* A "healthy" system at the same time effectively resists and copes with qualitative change; its flexibility in dealing with the unexpected makes life possible on both sides of the boundary separating two stable regimes.
12. Evolutionary process implies *openness as self-transcendence* and thus *imperfection, courage, and uncertainty*—not the deterministic perfection, static security, and certainty inherent in the ideals of the traditional structure-oriented Western world view.

Human design, undertaking to align with evolutionary emergence and to follow and ameliorate the above principles, will reverse many courses of action prescribed by contemporary planning and stabilization paradigms. In particular, as many chapters in this volume bear out, we seem to have arrived at a new evolutionary threshold, marked by a novel and unique task beyond that of being open to intermittent cultural transformations and designing an "infrastructure" for each regime. This new task amounts to the conscious design of a life of quasi-continuous qualitative change, pluralistic culture, uncertainty, variability, and high fluctuation—of coping with the unexpected through a "dynamic balance" which is to be found, not in any stationary regime, but in a metaregime of "fluidity in transformation." In this way, a transcultural sense of mankind may develop. Going *with* the fluctuations also seems to become the best way of "defusing" them when they tend to get so big as to threaten the long-term viability of an evolving system—as they certainly do in today's highly metastable (resilient) society, which has itself prepared the means for its own extinction.

This volume may be regarded as a challenge to develop a new understanding of human systems as an expression of *life*, a challenge to go beyond the familiar focus on adaptation and stabilization in order to develop a paradigm of self-realization through self-transcendence.

ISBN 0-201-03438-7/0-201-03439-5pbk.

Self-transcendent systems will then appear as an integral aspect of evolution—in fact, as that aspect of evolution which provides a vehicle for qualitative change and thus ensures its continuity. Such a new paradigm may be expected to develop into a new *science of connectedness*, which will also become the viable core for a new *science of humanity*.

Erich Jantsch
Berkeley, California

P.S. Shortly after he sent his "Concluding Remarks," Conrad Waddington died (26 September 1975). For those who knew him personally or who were through publications touched by his lifelong fascination with the creative dynamics of life, the generation of a whole world of forms through the interaction between organism and environment, it will be more than accidental that his last book project involved him in helping with a volume that probes into human life as it appears in this vast perspective. Extending the evolutionary paradigm to embrace the human sphere, celebrating life across all levels at which it becomes manifest, this volume now stands in memory of Conrad Waddington's life. A line from a French poem keeps coming back to me: "Et la Vie l'emporta . . ." (And Life took him away).

E. J.

ISBN 0-201-03438-7/0-201-03439-5pbk.

Part I

Self-Transcendence: New Light on the Evolutionary Paradigm

Man follows the ways of the Earth,
The Earth follows the ways of Heaven,
Heaven follows the ways of Tao,
Tao follows its own way.

Lao-Tzu, *Tao Teh Ching,* 25

Evolution, or order of process, is more than just a paradigm for the biological domain; it is a view of how a *totality* that hangs together in all of its interactive processes moves. This dynamic totality spans a vast spectrum from subatomic processes to social and further on to noetic (mental and psychic) processes. Across this spectrum, interactive processes spawn systems which in turn spawn new processes—in an evolutionary view, process and structure become complementary aspects of the same evolving totality. But it is not sufficient to characterize these systems simply as open, adaptive, nonequilibrium, or learning systems; they are all that and more: they are *self-transcendent,* which means that they are capable of representing themselves and therefore also of transforming themselves. Self-transcendent systems are evolution's vehicle for qualitative change and thus ensure its continuity; evolution, in turn, maintains self-transcendent systems which can only exist in a world of interdependence. For self-transcendent systems, Being falls together with Becoming. In this perspective, life becomes a much broader concept than just survival, adaptation and homeostasis: it constitutes the

ISBN 0-201-03438-7/0-201-03439-5pbk.

Erich Jantsch and Conrad H. Waddington (eds.), *Evolution and Consciousness: Human Systems in Transition.*

creative joy of reaching out, of risking and winning, of differentiating and forming new relations at many levels, of recognizing and expressing wholeness in every living system. Creativity becomes self-realization in a systemic context. In the evolutionary stream, we all carry and are carried at the same time.

E. J.

ISBN 0-201-03438-7/0-201-03439-5pbk.

1

Evolution in the Sub-human World

Conrad H. Waddington

Human evolution is a continuation of the evolutionary processes which went on before man appeared on the scene. Man's development of language as a means of communicating information and instructions from one generation to the next has, of course, provided him with an enormously powerful mechanism of evolution which is only very dimly foreshadowed by the incomparably less efficient means of cultural communication available to living things during earlier ages. But although this change of emphasis is so great that it may be considered to bring about a qualitative alteration between human evolution and that of other animals, nevertheless there is still some continuity in the character of evolutionary processes; and on examination the alteration does not turn out to be quite so drastic as it appeared in the light of the conventional, and by now old-fashioned, notions about animal evolution.

There have been several rather profound changes in our theories of biological evolution since Darwin's time. When *The Origin of Species* appeared, there was no real understanding of the process of heredity. When Darwin spoke of random variations, he meant variations in the living creatures as one observes them. He took it, and in fact demonstrated, that some of these variations would reappear in their offspring, and to that extent were hereditary. But this did not always happen, and he had little or no notion when it did occur or how. Gradually, something like a theory of heredity developed, under the auspices of statisticians such as Francis Galton and Karl Pearson, from studies on the resemblances and differences between brothers and sisters, parents and children, grandparents and grandchildren, first cousins, and so on. This statistical theory was never very profound or satisfactory, and was eventually swept away by the rediscovered Mendelian system of heredity carried by separated hereditary factors or genes. These genes were shown occasionally to change suddenly from one form to another, with no apparent rhyme or reason. To the early Mendelians, such as William Bateson, de Vries, and others, biological evolution seemed to depend on

ISBN 0-201-03438-7/0-201-03439-5pbk.

Erich Jantsch and Conrad H. Waddington (eds.), *Evolution and Consciousness: Human Systems in Transition.*

the appearance of single new genes, which were produced by a process which was "random," in the sense that it was in no apparent way dependent on the environment, but yet produced an effect that increased the probability that its carrier would leave offspring in that environment. These views were gradually developed, and finally given some mathematical form around 1930, which was the heyday of the idea that evolution consists of nothing more than random mutation and natural selection, or "chance and necessity" as Jacques Monod has recently called it.

By 1940, however, two new trends of thought began to be developed, which fairly soon radically altered the whole picture. The crucial point of the first, which was largely due to the experimental and field studies of Dobzhansky, was that it is not the case that in a population of organisms in nature most members have almost the same genes (the wild type) with only a few rare variants, so that the population has no resources for evolutionary change, until by chance a new suitable gene turns up. On the contrary, in a population every individual differs from every other in a large number of genes, many, of course, of very small effect. Thus the realistic picture of a natural population is to consider that it shares a large pool of highly diverse genes, each individual containing a particular sample drawn out of this pool. If for some reason an evolutionary alteration is called for, genetic potentialities to meet the new needs are almost certain to be already available within the pool; although to meet the need fully, many genes with minor effects tending in the right direction may have to be brought together and concentrated. According to this picture, the role of mutation is not to be a fundamental constraint, so that evolution is always waiting on the appearance of the appropriate mutation. Rather what random mutation does is to feed into the pool a continuous supply of new minor variants, so that there is always a rich mix available, from which natural selection can pick out those which suit its purposes at the time. One might say that natural mutation is providing the aggregate for forming concrete, and it is from the concrete blocks, rather than from the individual pebbles within them, that the evolving animal is constructed.

The other new idea has even more radical consequences. It reintroduces into evolutionary thinking the concept of the organism as it appears in nature, as Darwin saw it, that is, as a phenotype or developed individual. The earlier evolutionary theories had reduced the organism to its hereditary constitution, its genes—whether a few genes, as in the early years of the century, or the whole population of a gene pool. But, of course, the test of natural selection—the test of leaving offspring to the next generation—is not applied directly to the genes themselves, but to the organisms which carry those genes. If this notion is applied seri-

ISBN 0-201-03438-7/0-201-03439-5pbk.

ously and consistently, very radical changes in outlook become inevitable. In the first place, the development of an animal under the influence of its genes is obviously an intricate and carefully controlled process, which normally leads to a rather standard and invariant end result. A mutation of one or even several of the genes concerned can either disrupt the process completely, so that the animal dies, or it can produce effects only of a restricted character. If, for instance, you have a set of machine tools for producing a reciprocating internal combustion engine, changes in these tools can alter the character of the engine—the diameters of the cylinders, the stroke of the pistons, the clearances of the valves, and so on; but they can scarcely at a blow start producing a turbine or a jet. Similarly, with a developing biological organism, mutation is anything but omnipotent to produce changes in any conceivable direction.

An even more radical rethink is called for by the consideration that the environment in which development occurs has an effect on the character of the phenotype which will be produced. This introduces a very remarkable type of logical indeterminacy into the whole system. One and the same set of genes may produce different phenotypes if they develop in different environments; or again, some differences in genotype may fail to come to expression when they develop in particular environments, so that identical phenotypes are produced. There is, therefore, no one-to-one correlation between the genotype which will be inherited, and the phenotype on which natural selection will occur. The theory that one can reduce natural selection to the simple notion of "necessity" is thus completely untenable.

Another whole range of problems—perhaps the most interesting ones of all—arise when one inquires what determines the character of natural selection. There may be some factors in it which are in effect unavoidable for a particular animal. An elephant born in the middle of Africa must, whether he likes it or not, put up with a certain regime of temperature, wind, available plant food, and so on, and can do little more about it than seeking out the shade if he wishes to. A beetle in the same place, however, can do much more to select his appropriate local climate, for instance by burrowing into the soil or under the bark of a tree. In fact, a surprisingly large amount of the environment which exerts natural selection on an animal is the more or less direct result of the animal's own behavior. Quite often the animal has the choice that if he does not like it here he can go someplace else. Again, it is often the animal's behavior which decides whether he is selected for his ability to run away and escape from a predator, like a horse or an antelope, or for his ability to stand his ground and fight it off, like a buffalo; and, of course, the behavior which an animal will exhibit now must have been

ISBN 0-201-03438-7/0-201-03439-5pbk.

the evolutionary result of natural selection operating on his ancestors according to how they behaved in earlier periods. We have a typical cybernetic circularity of causation.

These considerations are not merely theoretical developments; they have suggested experiments which have revealed a new and very important mechanism of evolution. One of the most important aspects of evolution, indeed probably the most important of all, is the problem raised by the very precise manner in which organisms are adapted to the circumstances of their life. These were the phenomena from which it can be argued that the world has been created by an intelligent designer; and it was mainly because Darwin's theory offered an alternative to this argument that it aroused such enormous interest in the whole civilized world.

Now it is well known that many organisms can become adapted during their lifetime to the particular demands of the environment; if they use their muscles, they become stronger; if they are called upon to exercise a certain skill, they improve at it. But in spite of great efforts, no one has ever been able to demonstrate that such effects are inherited and passed on to the offspring, as Lamarck for instance has suggested they should be. The orthodox biology of the first few decades of this century therefore rejected them as totally irrelevant to evolution. The new point of view sketched above suggests one should look at the matter in a new light. Natural selection acts on phenotypes, which may have become adapted during their lifetime by physiological processes to the particular stresses they have had to meet. The selection will act on a population of organisms which have been subjected to the stress, and there is almost bound to be some genetic variation in the capacity of different individuals to respond appropriately to that stress. Selection will therefore tend to increase the frequency in later populations of genes conferring ability to respond adequately and adaptively. This in itself will merely mean that the capacity to respond adaptively to the stress will increase in later generations.

The situation becomes more interesting when one takes account of another aspect of the open-ended, nonlinear character of living organisms. Development from egg toward adult involves a good deal of cybernetic control, of such a kind that there is a tendency for the normal adult condition to be attained even when circumstances are far from normal; unusual stresses have to be quite severe before they succeed in producing developmental changes. If they do so, and the changes are adaptively useful and therefore improved by natural selection, the systems undergoing evolution will still be ones in which there will be cybernetic controls tending to produce a resistance to change. Suppose, then, that in the face of this resistance natural selection has

ISBN 0-201-03438-7/0-201-03439-5pbk.

nevertheless succeeded in evolving a type of organism with a new and effective adaptation to some novel condition, and that this novel adaptation will itself be somewhat stable and difficult to alter. Even if the conditions of stress which initially precipitated it are removed or disappear, the adaptation may persist for quite a number of generations. It may indeed need quite a strong selection, definitely against it, before it can be got rid of.

An adaptive change, originally induced by some environmental stress, may, after many generations of selection, form part of the hereditary endowment of a later population, appearing even in the absence of the stress. This process, which was at first postulated on purely theoretical grounds, has been amply demonstrated in experiments in the laboratory, and has been shown to occur in nature. It is an evolutionary mechanism which was not contemplated during the first 40 years of this century, when the orthodox classical theory of Mendelian evolution was being developed. It now provides an acceptable and scientifically convincing alternative explanation for all that class of phenomena for which in the past people have been tempted to invoke the Lamarckian "inheritance of acquired characters," but had reluctantly to restrain themselves, because they were convinced that the direct inheritance postulated by Lamarck does not in fact occur.

Once we consider evolution in terms of the selection of phenotypes which are produced by the development of a sample of genes drawn from a large gene pool, under the influence of an environment which is both selected by the organism and then selects the organism, we find ourselves forced to conclude that biological evolution, even at the subhuman level, is a matter of interlocking series of open-ended, cybernetic, or circular processes. In these, cultural transmission of information plays a relatively minor role, although it is probable that in some of the more highly evolved animals imitation of elders, for instance, has some influence on the selection of behavior strategies, such as escape from predators by running away among horses. Human evolution is, of course, characterized by enormously greater development of this type of sociogenetic mechanism, operating through the agency of language. The point I wish to make in these introductory remarks is that the biological process to which these human mechanisms are extensions is already one whose logical structure is far more subtle than a simple logical sequence of cause and effect.

ISBN 0-201-03438-7/0-201-03439-5pbk.

2

Openness as Self-Transcendence

Walter Pankow

1. INTRODUCTION

This volume attempts to describe the paradigm of open systems and to understand some of its consequences for human life. All living systems and all supersystems which are built from living systems are open systems. They are open with respect to matter, energy, and information which they exchange with their environment. This kind of openness I call *formal openness.* The introduction of this concept results in a vastly enriched understanding of living systems as compared with older concepts based on models of closed systems. In focus here is the dynamic structure, or in other words that basic characteristic of life which expresses itself in the paradox "continuity in change"; it may be sufficiently explained by the concept of formal openness.

In spite of this, formal openness is not sufficient for the understanding of living systems. The concept of openness—and this constitutes the principal thesis of this contribution—has to be expanded so as to become the concept of self-transcendence. But something quite unexpected happens in the process. I should like to illustrate this with a parable:

> Somebody draws a landscape with a pencil. In this way, a picture takes form consisting of lines, points, and gaps. By exploiting further the potential of various nuances of grey, the likeness of the image can be greatly enhanced. However, the interplay of different colors cannot be represented with the same pencil. The means of representation cannot be improved any further, but has to be questioned itself.

With reference to our problem this means that our representation (model, image) of the world—in our example the representation of living systems—depends not only on the world, but also on the means of representation, or in this case on language, which is the "organ" of con-

ISBN 0-201-03438-7/0-201-03439-5pbk.

Erich Jantsch and Conrad H. Waddington (eds.), *Evolution and Consciousness: Human Systems in Transition.*

sciousness. Formal openness may be described (represented) with the help of formal languages, but not self-transcendence. To describe self-transcendence, languages that are themselves self-transcending are needed. From the same consideration it follows that self-transcendence cannot be deduced in an *a priori* sense from the observation of objects (such as organisms or ecosystems), but only from an observation of observation, and therefore from the observation of one's own consciousness, in particular of one's *own* use of language. Only if I experience self-transcendence in my own thinking and speaking can I also recognize self-transcendence in the being and consciousness of other living systems. Only if my means of representation are themselves self-transcendent can I represent other self-transcendent objects; and my ultimate, most direct means of representation, which is also accessible to others, is my use of language. If I discover self-transcendence in this realm, which belongs to me in a most personal way, I can also recognize it in the outer world by means of sensitive feeling, i.e., through feeling entering and recognizing outer world.* The communication of such a recognition is profession (a professor, a teacher of science, is somebody who professes). Furthermore, I can tell others of my discovery only if they, too, know the personal experience of self-transcendence. Communication about self-transcendence does not work out of necessity because it presupposes the experience of self-transcendence. Certainly such an experience may be assumed for all humans. On the other hand, communication about self-transcendence is attacked, or altogether denied, to a greater extent today than ever before in history. It is interesting to note that this attacking is done with particular vigor where, originally, society had intended to further communication the most, namely, at the university.

This chapter attempts to show the extent to which self-transcendence is the key to understanding man and his societal systems as well as nonhuman living beings and ecosystems. Thus, self-transcendence becomes the common interdisciplinary beginning and end for the humanistic as well as the natural sciences. Interdisciplinarity through self-transcendence does not require the formalization of disciplines, but unifies the disciplines while preserving the variety of their ways of thinking and speaking (points or angles of view). In this contribution, I propose a specific concept of formal and self-transcending ways of thinking; it is, of course, based on my personal way of looking at things. I proceed to elaborate this concept by applying it to the problem of formal thinking in science and subsequently by attempting to discuss the emergence of self-transcending thinking in the child. In

*Homologous representation, discussed later

ISBN 0-201-03438-7/0-201-03439-5pbk.

doing so, I am bringing two groups of problems to the foreground: (1) the relationship between perception and behavior (being) which further develops in the human domain to become a relationship between thinking and doing (consciousness); and (2) the relationship between parts and wholes, individuals and society. These relationships can only exist simultaneously; they are prerequisites for each other.

2. GESTALT AND FORM

The task of language is to re-create or represent a perceived world in our consciousness (Zastrau, 1975). From now on we shall also use for the perceived world such synonyms as phenomenon, object, or system. But first I define two types of language:

1. *Natural languages* or everyday languages are defined by their capability of representing not only objects, but also themselves (Watzlawik et al., 1967). I call this characteristic *self-transcendence*, logical openness (discussed later), or just *openness*. A self-transcendent system I call *gestalt*. From the nature of self-transcendence it follows that there cannot be any two gestalt systems which are absolutely identical. Therefore, gestalt systems are *individuals*. By definition, all natural languages are gestalt systems. Of course, not all gestalt systems are gestalt languages, but they may be adequately represented by gestalt languages. I call this kind of representation homologous representation. Thus, gestalt systems are related to each other through *homology*. We shall see that all living systems are gestalt systems. This statement cannot be proven but can only be experienced directly. In this way, I expand the notion of life in such a way that it coincides with the notion of gestalt.
2. The other type of language embraces the *formal languages*. I define them by their inability to represent themselves. A formal language cannot make any statement referring to itself. Russell and Gödel (1931) have shown that this definition holds for all logical languages (quoted in Watzlawik et al., 1967). Thus we may characterize the property of non-self-transcendence as *logically isolated*, or simply *logical*. I call a logically isolated system a *form*. Thus, all formal languages are also formal systems. We shall discuss later the question whether there are other formal systems besides formal languages. Formal systems are related to each other through *analogy*. The adequate language representation of form is an analogous representation.

A statement *about* a statement is, for example, metacommunication (or command) about a report. Command and report are related to each other through a *hierarchy*. Meta-information, or instruction, about a logical language can only be given in another language, which may again be logical. Generally speaking, meta-information is given in everyday

ISBN 0-201-03438-7/0-201-03439-5pbk.

language. Examples from cybernetics for the use of logical meta-information may be found in the building of control hierarchies and computer program hierarchies. However, the ultimate meta-information is always formulated in everyday language. Mathematical textbooks, too, have to be cast in terms of everyday language.

The boundary separating here information and meta-information (instruction) lies *between* both and isolates both from each other in a logical way. I call this type of boundary a logical or *formal boundary.* Report and command relate to each other in the same way as do means and ends. This relationship is also characterized by *unilateral disposability*. We may also call this relationship an *analogous hierarchy.* A logical statement which makes a metastatement about itself is by definition *logically unresolvable* or *meaningless.* If, in spite of that, such statements are made, they appear in two types, namely, self-confirmation or *tautology* and self-renunciation or *paradox.*

In contrast, if we are dealing with natural languages, we may give a command in the same language as the report. It is thus that each statement in a natural language is always at the same time a meta-information. It is impossible to talk without at the same time talking "between the lines," in other words, without interpreting what is being said. We may also say that for each statement there is at the same time an *aspect of contents* (information, report) and an *aspect of relation* (instruction, command) (Watzlawik et al., 1967, pp. 51ff.). Therefore, each statement comprises complementary aspects. It is itself a *complementarity.* The boundary between report and command, information and instruction, does not separate the two, but is an expression of the communication itself. We may even say that this kind of boundary unites rather than separates. If we call this boundary a *natural boundary* and the hierarchy of report and command a *natural hierarchy*, then we may link the following synonymous notions:

Natural boundary; natural hierarchy; complementarity; self-transcendence; gestalt; individual.

The statement that all living systems are gestalt systems may now be understood in clearer terms because, viewed from a certain angle, living systems are membranes and skins. Their essence is not so much *formal* openness but *logical* openness, namely, self-transcendence.

A statement in everyday language which makes a metastatement about itself is *not* meaningless, but resolvable. However, this resolvableness is not unilaterally disposable, but requires *humor* and *confidence* (Watzlawik et al., 1967). We have already seen that each statement interprets itself, gives itself a specific quality, but that also the special cases of tautology and paradox become resolvable by means of humor.

ISBN 0-201-03438-7/0-201-03439-5pbk.

(For example, a mailman is told to fetch mail from the post office only for those people who do not fetch their mail themselves. Only a logician will believe that this order is unresolvable. Because, indeed, the mailman is allowed to fetch his own mail only if he does not fetch it, and vice versa.)

Awareness and Communication

Each re-creation of perception through language is *interpretation* of perception, for one's own consciousness as well as for that of others. *Awareness* and *communication* are therefore also complementary. I am aware of myself and of my environment only to that extent that I am able to communicate and make myself understood. Hora (1959, quoted in Watzlawik et al., 1967, p. 36) has formulated this in the following way: "To understand himself, man needs to be understood by another. To be understood by another, he needs to understand the other." The two questions are really the same: How can an individual become aware of himself and his environment, and how can a partner understand a communication by means of language? The answer is given by the self-transcendence of individuals and of language.

Self-transcendence means the capability to change one's own point of view, and therefore the capability to view a situation in a new light, or, one might say, the ability to jump over one's own shadow. The symbol for such a capability is Baron Münchhausen, who pulled himself up by his own hair. Thus, self-transcendence makes individual and mutual understanding possible by means of the very capability of including the new in the old, as well as the individual in the common, because individual and collective consciousness have developed, and still develop, in mutual interdependence. A natural language, therefore, also has two more complementary characteristics: It is at the same time *pre*established, objective consciousness (prejudgment) and *newly* designed, subjective consciousness (judgment). The prejudgment, so to speak, carries the judgment, is modulated and designed by the latter, but always remains open to the new. Self-transcendence is at the same time the cause for the differences in the points of view taken by different individuals as well as within the individuals themselves (which undergo continuous change) and the prerequisite for overcoming these differences through understanding without abolishing variety. Only an open language is capable of linking different points of view. *Only openness is capable of understanding openness. Only gestalt is capable of understanding gestalt.* If, as was said earlier, it is impossible to talk without simultaneously talking "between the lines," we recognize now that this must

ISBN 0-201-03438-7/0-201-03439-5pbk.

necessarily be so: Line and space between the lines are inseparable and complementary. Understanding is not unilaterally disposable, it requires confidence. In this way, awareness and understanding become creative, induce movement and change of position, and thereby make individuals change and evolve.

Time

Being is becoming. Self-transcendent systems are time-generating systems. Being and time are complementary aspects of gestalt. Gestalt has no origin; it is its own origin.

3. PHILOSOPHY AND MATHEMATICS

So far we have discussed two types of language, natural language and formal language. In the same way, we may speak of two types of science, the philosophical sciences and mathematics. Philosophy is concerned with recognition and poses the basic question "What is?," together with its elaborations "Why?" and "How?" In this respect, all nonmathematical sciences, natural science as well as the humanities, are alike. All nonmathematical sciences claim to exist not just for their own sake, but for the sake of understanding our world as we personally experience it *directly*. In this respect they are philosophy because they are concerned with the world which concerns us, which interests us, and in which we share. Only one science excepts itself from this purpose: mathematics. It is self-sufficient, does not look beyond itself, and builds its own formal world.[1] It does not need to justify itself because the other sciences are only too eager to procure its services. Mathematics supplies the sciences with formal models which are then used as analogies for the observations made by the sciences. But only form can

[1] These remarks refer to the Western formalistic use of mathematics. The qualitative view held by other cultures is coming to the fore again in the context of modern physics and Jungian psychology: "Consequently, it is not only the parallelism of concepts (to which Bohr and Pauli have both drawn attention) which nowadays draws physics and psychology together, but more significantly the psychic dynamics of the concept of number as an archetypal actuality appearing in its 'transgressive' aspect in the realm of matter. It preconsciously orders both psychic thought processes *and* the manifestations of material reality. As the active ordering factor, it represents the essence of what we generally term 'mind' " (Franz, 1974, p. 53). Chapter 9 by Marney and Schmidt seems to suggest a similar view of mathematics as an evolving physical/psychic actuality. (Comment by E. J., editor)

ISBN 0-201-03438-7/0-201-03439-5pbk.

be modeled adequately by analogies. Therefore, we are thrown back to the initial question: "Where, outside mathematics, does form exist?" I believe we can answer: For sure, under certain conditions, in the world investigated by physics. There, too, may be found the key to technology. Furthermore, we have to ask: "What happens to the observed systems and to us if we apply mathematical models?" Before exploring these problems of physics, it is necessary to discuss in greater detail the properties of form and the opportunities offered by mathematics.

Mathematics

By definition, formal languages are incapable of recognizing themselves and, as we have seen, are therefore also incapable of recognizing any other position. Thus, a formal language does not link positions, but constitutes itself a fixed position. It does not see the world beyond itself, only the world within itself. It is not, like philosophy, an open eye to the outer world, but a closed eye which may also be called a logically isolated eye. Mathematics is the science of form. In the following, we shall use the notions "formal," "logical," and "mathematical" as synonyms. As a consequence of logical isolation, the boundaries of a form cut through the complementary aspects of statement and metastatement. In this way, the boundaries separate elements from relations between these elements. In mathematics, the elements are called *variables* and the relations between them *functions.*[2] From logical isolation it follows furthermore that *time* itself is logically isolated and may therefore be viewed as an independent variable. With its help, any form may now be represented in a timeless way because time itself may be represented as the time coordinate in a timeless way. For example, time relations expressed through graphs on a sheet of paper do not change once the curves have been drawn. These properties of form may also be expressed through the identity principle of logic: Variables remain *identical* with themselves from a qualitative point of view: they can change only in a quantitative way. This is expressed in such a way that one says, they can assume different values.

In this way, a mathematical system consists on the one hand of elements, also called variables, and on the other of relations between variables, also called functions, which constrain the possibilities of change open to the variables. A system *state* is defined as a set of variables with given values. A change of state is a change in the values of variables. A mathematical *structure*, then, is the description of the dy-

[2] I do not intend to give here a description of the whole field of mathematics, but only of the mathematical tools for the natural scientist.

ISBN 0-201-03438-7/0-201-03439-5pbk.

namics of the system—in other words, of the sequence of states it goes through.

If the functions permit only one direction of change—which means the system has only one degree of freedom—one speaks of *deterministic* or reversible systems. If a system is not subject to functional constraints of the elements, we have, with a large number of elements, an indefinite number of degrees of freedom. In this case, one speaks of *stochastic* or irreversible systems, which have many more possibilities than they are capable of realizing. System states are then no longer defined in a microscopic way, that is, for each variable and each value, but in a macroscopic way by classes grouping microscopic states. The values of the variables are replaced by probabilistic values of classes. In such a way it now becomes possible to represent the dynamics of stochastic systems too in a timeless way.

The most general type of mathematical systems consists of a large number of elements which are only partially constrained by functions. In this way, the systems exhibit deterministic as well as stochastic features. A description of their dynamics is possible only for special cases. Systematic inquiry into their behavior yields a vast variety of dynamic structures. In particular, the introduction of feedback between the elements leads to a novel type of structure which gives rise to an enormous variety of possible developments: dissipative structures (see also Chapter 5 by Prigogine). To the extent that inquiries into special cases are carried out with a view toward contributing to a general theory of mathematical systems, they go today under the name of systems science. A common characteristic of all of them is the use of computers, because this seems to be the only way to cope with the type of mathematical complexity involved. The computer is also becoming a factor in linking mathematics and physics; yet it cannot bridge the gap between mathematics and philosophy.

Physics

We shall now deal with the question whether form exists outside mathematics. Einstein reportedly considered it miraculous that mathematics is applicable to physics at all. After what we have said earlier, this applicability is indeed anything but self-evident. It is a pity that most physicists and, more importantly, most other scientists as well, avoid reflecting on this miracle. If Galileo appeals to scientists to measure whatever can be measured, it would be only rational to ask for the criteria of success of measurement. In particular, we ought to know where to obtain an adequate measure and when measurement can really be

ISBN 0-201-03438-7/0-201-03439-5pbk.

justified. In a formal language, this question cannot be posed because form cannot recognize anything beyond itself, not even other form, but has to presuppose it. Only a natural language can recognize the outer world, form as well as gestalt.

Let us look first at the ways in which measurement is carried out. A setup for measurement is called an experiment (see, in particular, Müller, 1972). All experiments are aimed at timeless (formal), reproducible natural laws which not only hold under the conditions set for the experiment but—and this is the real purpose—can also explain a directly observable phenomenon *outside* the experiment. We demand of measuring devices that they constitute analogue models of mathematical variables which can change only in their quantitative value. Therefore, they are not supposed to change their quality during the measurement. Whether the measuring devices actually used fulfill this requirement or not has to be presumed at first. The arrangement of a measuring device for the *indirect* observation of a directly observed phenomenon is called *operationalization* or *preparation* of the phenomenon. As we shall see later, each direct perception probably requires change in quality. We are blind to mere quantitative change. But the experimental preparation involves the *screening out of quality* and the setting up of a mathematical system which consists of variables in the form of measuring devices. What we are now looking for are the relations between the variables, namely, the functions. If these relations contain more than two variables, mathematical systems become complex and experiments are therefore usually carried out in such a way that only one degree of freedom comes into play. This means that only two variables are taken into account, one of which—the so-called independent variable (in many cases time)—is set by the scientist, and the second one, the dependent variable, is being observed. It is then important that the phenomenon be prepared in such a way that there are no other degrees of freedom besides the ones meant to be observed. All other variables which the system may have must therefore be kept constant. This procedure is called the setting of *boundary conditions.* Since normally it is not known which boundary conditions are relevant, as many as possible are usually controlled as extensively as possible. This is why we build laboratories.

In this way, therefore, a phenomenon becomes logically isolated and the only openness is the preset number of degrees of freedom. Thus we have transformed the direct phenomenon into a mathematical system: *If* we are successful in imposing logical isolation, *then* the experiment will yield the mathematical functions. Therefore, the gain from an experiment is an *"if/then" structure.* If I prepare the phenomenon in a specific way, then I obtain specific mathematical functions. Although

ISBN 0-201-03438-7/0-201-03439-5pbk.

this constitutes a law, it is far from being a natural law because the latter would require that such an if/then structure point beyond the particular setting of the experiment.

As a next step one has to try to investigate all boundary conditions relevant to the law indicated by the experiment. But for this, one has to set up new experiments which introduce former boundary conditions as new variables. If the result is contradictory to old laws, this may be interpreted as an indication of the relevance of the investigated variables. By also including these new variables in the mathematical system, contradictions may frequently be resolved by finding laws of laws which embrace the old laws as special cases. In this way, by generating contradictions (thesis and antithesis) and their removal (synthesis), physics proceeds to include more and more of the outer world by including an increasing number of boundary conditions in the experiment (i.e., in the mathematical system). This, however, means that physics transforms increasing portions of the outer world into a laboratory world.

There is one realm of the environment whose mathematical preparation always leads to deterministic and stochastic models and for which it is always possible to construct internally consistent multivariate total systems on the basis of experiments with only two variables and one degree of freedom. Such mathematical systems are called linear systems. Our initial supposition that measuring devices may be used as mathematical variables has now been proven *a posteriori* through internal consistency and remains confirmed as long as all emerging contradictions can be formally overcome.

In this way, the step from direct observation to a prepared situation is being continuously postponed, but it remains the crucial characteristic of physics viewed as a philosophical science. To recognize a law as a natural law, we need its confirmation in an unprepared outer world by many direct observers. Above all it is in technology that physical laws turn out to be natural laws.

Therefore, arranging formally consistent systems is *not* the only task of physics because people do not live in laboratories. However, this formal task is so huge that many physicists never need to see the light of day. As soon as a science observes the world *only* by means of instruments and thereby cuts the connection to any direct observation of phenomena outside experiments, it becomes blind to the outer world in the same way as mathematics. Statements about the outer world are then logically irresolvable, that is, tautological. To the extent that science becomes tautological without knowing it and therefore insists on its claim to explain the outer world, it will try to transform the outer world according to its own image and in this way it will become a *self-*

ISBN 0-201-03438-7/0-201-03439-5pbk.

fulfilling prophecy. Its successes are then no longer successes in explaining an observed world, but successes in transforming an unobserved world into the preestablished form of scientific inquiry. The explanation is no longer geared to the phenomenon, but the phenomenon is adapted to an already existing explanation. What I know already, I do not need to observe.

Perhaps quantum mechanics has brought some improvement in this respect, because there the experimental preparation generates formal inconsistencies which cannot be overcome in any formal way and remain as complementarities in Bohr's sense. Here we again need the direct observer to unite what can no longer be brought together formally.

More recently, physics is also concerning itself with the description of phenomena which cannot be represented by linear compositions of experiments with only two variables. To this domain belong all kinds of dissipative structures found in natural processes, and in particular in biochemistry (see Chapters 5 by Prigogine and 6 by Abraham). To cope with them, the first step again involves an assumption about the variables. But now it is assumed that variables may enter three, instead of two, relations. In this way, one arrives at a more general type of mathematical system characterized by a vast number of degrees of freedom and an enormous variety of possible dynamic structures. Such mathematical systems are also known as nonlinear or holistic systems. But now we encounter insurmountable problems of measurement. Therefore, we reverse the direction of inquiry. We no longer proceed from measurement to mathematical system, but from an assumed mathematical structure to vindication by measurement. This procedure is called simulation. It is no longer possible to construct the mathematical system step by step by adding new variables, because the system can only exist as a whole. This is the reason for its being called holistic. But even extremely simple assumptions lead here to tremendously complicated dynamic structures, so that they can be represented only with the help of computers. For purposes of evaluation, linear measuring instruments may be used because the macrobehavior of the simulation model may be assumed to be linear within certain limits.

In this new approach taken by physics, it now becomes essential that the experiment give up its claim to unlimited disposition over form. Phenomena are again observed primarily in a direct way. It is no longer possible to prepare the physical phenomenon in such a way that it can be represented by a microscopically adequate mathematical system consisting of analogous variables and functions. To attempt such a preparation would mean to destroy the phenomenon. There is no longer any experimental arrangement by which the "outer world" (the physical phenomenon) can be brought to match the "inner world" (the mathematical model). Therefore, the relationship between elements and

ISBN 0-201-03438-7/0-201-03439-5pbk.

structure, parts and whole, is indeterminate (Pankow, 1975). This indeterminacy constitutes an aspect of formal openness and by no means of self-transcendence because the dynamic structure is based on a timeless mathematical system. Thus, form in itself may already be so complicated that it can be modeled only in approximative ways.

Now our previous question, whether form exists outside mathematics, finds an affirmative answer. But this answer can only be given by testing experimentally derived and corroborated laws in the context of direct experience.

Let us now turn back again to man, that creature which is capable of disposing over form. The development of language, the organ of consciousness, goes hand in hand with the development of technology. Not only is language a conscious re-creation of our world of experience; the world of experience is also a concrete representation of language. Thinking and acting become one; they constitute a complementarity. According to Kant, the hand is the outer brain of man.

4. CONSCIOUSNESS: THE UNITY OF THINKING AND ACTING

There are various ways in which we can experience the unity of thinking and acting. For each of these ways we have different names. We already encountered a particularly striking example when we discussed scientific experimentation. Breaking up a system into single identifiable courses of action simultaneously constitutes the creation of a language, namely, a technical jargon. Let us now consider how a baby *learns.* In doing so, we are not only concerned with a certain aspect, namely, the contents of the description, but also with its meta-aspect, namely, *how* language speaks and how the words point to their own origin.

Not the disciple, but the baby is the learning creature *par excellence.* The task posed to the baby by mere birth is unimaginable. How is it supposed to cope with the new environment? The most difficult task is posed right at the beginning of life. To the same extent as the environment is new to the baby, the baby is also new to itself. The two basic questions "Who am I?" and "Who are you?" constitute his existential problem. The senses are not sufficiently prefigured. They develop together with the relations to the new environment. Not only does the structure of the organs develop ahead of functions, but always as a function.[3] Nevertheless the baby is anything but helpless. It mas-

[3] We have discussed self-transcendence in our own *consciousness* as complementarity of report and command. Starting with this insight, we are now in a position to refind a homologous coupling at the level of *being,* namely, the coupling of structure and function (see also Chapters 4 by Holling and 5 by Prigogine).

ISBN 0-201-03438-7/0-201-03439-5pbk.

ters its task not because it is prepared for it but, on the contrary, because it is open, because it is capable of learning, or even better, because it becomes ever better in learning to learn. Let us see how relations with the environment develop.

Perception

The starting point is the mouth. For the baby, it is the new umbilical cord, the link to the mother and the outer world in general. This link is the experienced unity of the complementarity of *sense* as feeling and as order. This perception itself is a unity of feeling and behaving, too. Perception requires action—in the case of the mouth, rhythmic sucking. Thus, perception is also the pulsation of active and passive behavior, of doing and being done to. To have is to gain. The pulsation is an expression of a continuous change of position, in other words, of self-transcendence, as I called it earlier. In this way, in the beginning of a development a complementarity of various ways of being becomes evident by synonymity: Life; being; sense; perception; behavior; time.

Attention and interest become stimulated only when there is in the environment a minimum of ambiguity, secrecy, fuzziness, and indeterminacy. *Indeterminacy* constitutes a challenge to self-awareness. As man brings order to his environment, he becomes aware of himself. The challenge posed to his interest by indeterminacy is vital. No creature can bear to bear nothing (Kükelhaus, 1974). Over and over again we notice the basic characteristic of self-transcendent systems, namely, being as becoming, the unity of existence and time. Living systems are therefore time-generating; they weave, so to speak, a pattern of time threads. The question, what was before life, can no longer be posed in this perspective: What was before time? Here, the preposterous preparation of time by means of clocks becomes evident.

During their further development, the organs which were initially oriented toward the mouth assume increasing independence. We might say they receive their proper capability of perception through the mouth. Thereby, the mouth itself is given new possibilities for unfoldment. In the rise of consciousness, the hand plays a key role.

Consciousness

The transition from prehension (grasping) to comprehension is symbolic of the emergence of consciousness. It is particularly worthwhile to note that this process is of a universal type and gives rise to the emergence of one self-transcendent system from another. It provides an ex-

ISBN 0-201-03438-7/0-201-03439-5pbk.

ample for Virchow's motto *Omne vivum ex vivo* (All life comes from life). Consciousness is a metaperception, the vindication of perception, the perception of perception. In this context, N. Tinbergen (quoted in Mislin, 1975) has made an important observation:

A baby crawls over a sand dune and gets scratched by a thistle. It stops crawling and, crawling backward, lightly touches the thistle again with its foot. Then it turns around and strokes the thistle with its hand. Up to now we have followed the perception of an object, triggered by the strange experience. But now something decisive happens. The baby crawls to another plant nearby which is not prickly and strokes it as well; in this way, it compares perceptions. Subsequently it returns to the thistle and *vindicates* the correctness of its distinct perceptions. With this corroboration, the perception "prickly" becomes disposable for the future, partly separated from the concrete situation on the beach. Perception becomes metaperception: in other words, it becomes consciousness. The baby has not only experienced, but has also explored this experience in two respects: *Where* does the experience "prickly" come from? *How* can it be repeated in the future? Thus, learning is not only recognition of repetition, but also something much more active: Experiences are prepared in the present with a view toward the future.

This example demonstrates the emergence of a concept, namely, the concept "prickly." But there is not yet any re-creation by language. Therefore, the concept emerges from an action. We may also say: The origin of the concept is an if/then relation, where the word "if" points to the context. The perception "prickly" at the level of the object refers only to the "then," whereas the metaperception refers to the "if." In discovering that "prickly" is conditioned, that it relates to a context, an origin, perception itself becomes an object, that is, it becomes conscious.[4] As the origin is being noted, the object becomes partially independent of the context and thereby disposable for the future. This constitutes a good example of Bacon's motto *Naturam parendo vincimus* (By obeying nature, we win). A concept thus is reified action.

Unfolding of Consciousness

Eventually the child learns many conditions which are connected with the experience "prickly." The context of experience expands. Thereby perception becomes more differentiated and detailed, a process which links up with the development of organs. We may also say: Conception unfolds through the inclusion of new experience. This, in turn, becomes

[4] In the German original, origin and object are expressed by means of a generic relation in the language: *Ur-Sache* vs. *Sache*.

ISBN 0-201-03438-7/0-201-03439-5pbk.

the basis for making the concept disposable in new situations. By tying ever more "ifs" to the "then," which also becomes more differentiated, and by making the ties ever more reproducible and certain, the action character of if/then structures retreats more and more in favor of an object character. Therefore, the increasing independence from the context does not relate to the separation of the "then" from the "if," but to the isolation of the if/then structure from the conditions of the "if," in other words, from the condition of the condition. However, this must not be carried too far, not to complete isolation. In such a case, the if/then structures would rigidify to form. The re-use of the concepts in new contexts would become impossible. Since the new, the unknown, the mysterious are constituents of any perception, complete isolation would equal becoming totally blind.

Again, we recognize that an excessively objective (formal) natural science is a contradiction in itself. The isolation of if/then structures from the condition of the condition we have also called operationalization, preparation, or screening out. Laboratories as realms of indirect experience are technical realizations of this process. Each conceptualization thus requires only partial screening. Laboratories are meaningful only under the condition that they do not obstruct the view of the world of experience obtained by science. Therefore, science must never remain exclusively within the laboratory.

So far I have discussed the unfolding of concepts without mentioning language. But the splendor of the unfolding of consciousness in the human world is possible only through re-creation of concepts by language. Only in this way can we become free in the flexible use of concepts. Disposition, then, no longer necessitates grasping; it is sufficient to *name*.

By means of conceptualization through language man is capable of forming images of his future, so that he can imagine various options and plans from which to select. Thus, man has the capability of expanding his future. History does not simply happen to him; he makes it by planning it. With this faculty he also gains the possibility of foreclosing his own future, for example by introducing rules for planning which allow only for extrapolations from past experience. This attitude is also called "keeping both feet on the ground," or "linear planning."

We have seen that conceptualization always has to leave a residue of openness (indeterminacy, fuzziness). Mutual comprehension, therefore, is not disposable in an unambiguous way as it would be in a formal language. This gap must be closed by *confidence*. Therefore, misunderstandings cannot be excluded. On the other hand, they are often an opportunity for expanding the horizon. It is the very challenge of the unknown from which the new rises.

ISBN 0-201-03438-7/0-201-03439-5pbk.

Unfolding of Time: History

The emergence of consciousness falls together with the emergence of *questioning.* Consciousness widens the margin between the factual and the possible. At the very moment in which man becomes aware of a lack of regularity, he also recognizes the possibilities inherent in this lack. In this way, a question becomes the trigger for a *wish* and at the same time the wish becomes the trigger for the question. The wish to understand and the wish to act independently are two sides of the same coin. Consciousness of time, that is, the perception of things which change and others which do not, generates the question "Why?" in the two versions "Where from?" and "What for?" Together with the possibility of action, that is, the perception of one's own individuality, the questions "How?" and "Whereby?" are raised. Man can only understand himself by understanding his environment, and he understands his environment only by transforming it actively into a world, *his* world. Therefore, man is the most open of all living beings, not only because he has the highest potential for development, but also because he recognizes this potential.

I have defined openness as self-transcendence. Furthermore, I have stated that all living beings are open, self-transcendent systems. In this perspective, I view the peculiar nature of man in the extreme enhancement of this basic characteristic of life. Man, so to speak, specializes in openness by creating a language consciousness which, in turn, is itself a self-transcendent system: In this way, one self-transcendent system gives rise to two, but the bond between them is never dissolved. As long as this bond does not break, the trend toward increasing independence of consciousness also enhances being. The same process also underlies the emergence of the time concept: From the original unity of time and being, a time aspect assumes ever greater independence. This time aspect now forms a background against which being unfolds. Out of a lived present emerges an experienced present by means of relating it to the past and the future. With this, man gains *history.* As long as the bond between time and being is always renewed and time does not become an independent variable of measured time, the trend toward independence of time enhances the options open to history. Thus, consciousness becomes time consciousness. It preserves its meaning by linking backward to being. This is what I understand under *re-ligio.* The dissolution of the *re-ligio* between being and consciousness would imply the destruction of self-transcendence, the blinding of consciousness, or even more precisely, the blinding of consciousness by its own brightness. Consciousness and time would then no longer be self-transcendence, but form. As a consequence, being would also be destroyed. In

ISBN 0-201-03438-7/0-201-03439-5pbk.

my view, such a process underlies our contemporary social and ecological problems.

5. INDIVIDUATION AND SOCIALIZATION

I have described in the preceding section the emergence of consciousness by using the example of an individual. But language as an organ of consciousness is at the same time eminently social. Talking plays the double role of becoming self-conscious and of communicating. In this way, an individual becomes conscious by using a preexisting language. This gives rise to enormous problems. How can a preexisting language fulfill the double task of self-representation and communication? Is it not true that structure and function develop jointly as complementarities? Is it not necessary for a *pre*existing language to *pre*scribe, that is, to manipulate self-representation and communication? In other words, how can new individuals grow into a preexisting community and thereby even enhance their individuality? How should we understand a kind of *socialization* which turns out to be a prerequisite for individuation?

In the development of consciousness, as it is reflected in the re-creation of the mother tongue, the joint expression through doing and speaking, with symbolic reference to simple and basic actions, plays a very important role. How dangerous it is, therefore, if everyday life is relegated to places to which there is no access to the unauthorized—and these are in all cases the children. Children are great imitators. By active and symbolic re-creation, they become certain of their perception, and thereby of themselves. Each re-invention of a set of means for representation is sought and refound in a variety of contexts. Identity can be found only by changing the context, and change becomes a challenge only in connection with identity. It now becomes obvious that the learning steps of a child always have to be connected with the total life process. The initial means of representation, like a tree, have to be capable of providing means for the unfoldment of ever more differentiated experiences. The link backward to the origin must never break. In such a re-created system of speaking and doing, sense as feeling and sense as order always merge into the sense of life. Even more, mutual understanding between different ages is thereby ensured.

This kind of learning and developing constitutes a free, spontaneous, and self-determined unfolding of the sense of life and at the same time the active adaptation of a new open system to an older one. The pacemaker is the new system itself. Any attempt to accelerate the

ISBN 0-201-03438-7/0-201-03439-5pbk.

process can only lead to its destruction, whether by means of untimely rules or by means of premature fulfilling of yet unexpressed wishes. The latter kind of acceleration is especially typical of our time. I hold that the conflict of mutual adaptation between the young and the old is possible also in freedom and that guidance does not have to be manipulative, and that this conflict is even a necessary constituent of freedom. The prerequisite for this, however, is a self-transcendent language.

Art and Technology

We have discussed man as a wishing creature. In this perspective, and to a much greater extent than any other creature, man does not live in an environment which is pregiven, but in a world which he creates himself. He creates his world as a mirror image of himself. To the extent that man creates the world, he creates himself, and vice versa. In this way, he comes to own the world. To own is not to buy something ready-made, but to express one's own self in what is owned. Self-representation may then mean two things: Becoming certain of oneself (individuation), and representation to the fellow-human (socialization). Thus, man is an artist and his means of representation are not restricted to his own body, but include the environment. Man not only lives, he acts out his life, he interprets and enhances it through gesture and mime, language and technology, both for himself and for others.

Technology is the capability of disposing over form. Form is free of self-purpose and therefore disposable to wishes from the outside. I call disposed form a *technological product.* If *art* is the way in which man becomes certain of himself, then the aim of technology is art.

Man views everything factual against the background of the possible, which in this way, becomes present. Facts which foreclose options do not exist for man. Man becomes a technologist by keeping form open in its disposability and by using it. In this way, he creates technology which is a means of representation at the disposition of the community. Thereby, knowledge keeps flowing from the individual to the community and back. This kind of art exists as long as the flow continues. Structure exists only in its function. Only by speaking a language do we maintain it as a language. No encyclopedia is capable of saving a language which is no longer spoken. The same holds for art. Only by creating and re-creating, seeking, finding, and refinding, does art remain present. Museums cannot preserve art. Only individuals relating to works of art can keep art alive. In this active relationship, the past is continuously being brought into the present. The present carries its total history within itself. History may be compared to a tree: The

ISBN 0-201-03438-7/0-201-03439-5pbk.

individuals of the present are the outer branches; the roots reach far down into the past; but in the present, it is the whole tree which is alive. Humanity, as we encounter it in the present, continuously actualizes total history. Individuals bring the potential of history as a whole into the present. Each individual does this in a new way. Individuals understand each other to the extent that they have the same history, that is, that they relate to the same branch of the same trunk. The tree of which we are speaking here has different names: history, language, or art. Even in its outermost buds, there is the potential of the tree as a whole. The tree does not paralyze the unfolding of the individuals, but refines it and provides guidance to the individual. Art enhances the potential for self-expression. Even the most complicated art always remains simple. Art does not constrain the freedom of the individual. It does not impose itself.

However, there is not only the feedback loop between individual knowledge and communal art. In this loop, technological products are continuously generated which, being form, fall out of time. They share in time only to the extent that their design incorporates individual expression, that they can be made present and thus are *im*perfect and pose a challenge to be re-created. Art requires of technology that the latter dispose over form only to such an extent that form remains disposable to art. If, for example, the product is a house, it must not unduly restrict the lives of its tenants.

We might ask why humans, who possess art, make life so difficult for themselves. They know of time, they can *fore*see, orient themselves toward the future. They can also design the world in which they live. How can it then be explained that they experience this potential not as joy, but as anxiety? Men want certainty, not hope. They cannot bear the uncertainty which is so basic to all life. They are hopeless. They do not develop a potential, but ruin it. But we know that a future which is certain is timeless and therefore form. What is it which makes man so hopeless that he experiences hope itself as anxiety and finds security only in death? Dostoyevsky expresses this dilemma most profoundly in his parable of the Grand Inquisitor in *The Brothers Karamazov.*

Division of Labor

We have seen that individual and society are not opposites, but constitute each other. Nevertheless, it remains puzzling how this mutual harmony can develop, because harmony is not a very dominant feature of our world society.

ISBN 0-201-03438-7/0-201-03439-5pbk.

Society may be characterized as a *system of divided labor.* However, we must be very careful not to block our view by applying common prejudices concerning the notions of labor, division, and system. I have already stated that humans tend to separate what belongs together and to link what does not belong together. In the issues concerning organized divided labor we encounter again the old problem of the relationship between the whole and its parts which we have already discussed in several aspects. But now, in this formulation, the connection with our present situation becomes even clearer.

In all the examples evoked earlier we encountered two types of systems: form and gestalt. These types also correspond to two ways of dividing. Formal division separates the two sides of complementarities and treats them as *identities.* Natural division separates from each other *units* which keep all their complementary properties. The separated units are *individuals,* and thus identical neither with each other nor with themselves; nor are they strange to each other. From this follow two possible ways of joining units:

1. *A priori:* Units are not completely separated, but *remain* linked. Our discussion of time has shown that this is always the case in some way. The link backward to the origin never breaks.
2. *A posteriori:* Units which stem from different branches of division join to form a *new* bond.

Both types of bonding always cooperate in a complementary fashion. Each unit enters bonds of both types in the *unfolding of the old through inclusion of the new.* Since many branches of division may be traced back to the same original units, it is often impossible to distinguish between old and new bonds. I prefer to leave open here the interesting question whether all branches lead back to the same unit.

Cell division and differentiation in the embryonic development of an organism are obvious examples of such a process of division and bonding, since in the cell we encounter the prototype of a living unit. But we have also gained a view of far wider scope comprising the emergence of consciousness from being, the unfolding of concepts in language, individuation and socialization. The new way is the mode of thinking encountered with human consciousness and therefore also the mode capable of representing life in the widest possible sense, that is, in the sense of a self-transcending gestalt.

The question "What is the relationship between a whole and its parts?," now receives the puzzling response: "Parts are themselves wholes." The inquiry into these relationships in all domains of life and their homologous representation constitutes the most wonderful task

ISBN 0-201-03438-7/0-201-03439-5pbk.

for science. As an example, I should like to mention here the phenomenon of sexuality, which as a prototype for the joining of the new and the old points far beyond biology to a general theory of gestalt. But for me, the most beautiful feature of science in this perspective is that it cannot be monopolized. It is a science which every human carries within himself, a science for conviviality.

REFERENCES

Franz, Marie-Louise von (1974). *Number and Time: Reflections Leading toward a Unification of Depth Psychology and Physics* (A. Dykes, transl.). Evanston, Ill.: Northwestern Univ. Press.

Gödel, K. (1931). *Monats. Math. Phys.*, 38, 173–198.

Hora, T. (1959). "Tao, Zen, and Existential Psychotherapy," *Psychologia*, 2, 236–242.

Kükelhaus, H. (1974). *Unmenschliche Architektur.* Cologne: Gaia.

Mislin, H. (1975). "Angeboren-umweltbezogenes Verhalten als Grundlage frühkindlicher Ethik," paper presented to the conference *Organismus und Technik*, Munich, 1975.

Müller, A. K. M. (1972). *Die präparierte Zeit: Der Mensch in der Krise seiner eigenen Zielsetzungen*, Kap. II, 1.2. Stuttgart: Radius.

Pankow, W. (1975). "Die dynamische Organisation ökologischer und sozialer Systeme," in *Technik für oder gegen den Menschen* (P. Fornallaz, ed.). Basel: Birkhäuser.

Watzlawik, P., Beavin, J. H., and Jackson, D. D. (1967). *Pragmatics of Human Communication.* New York: Norton.

Zastrau, A. (1975). "Organismus als Sprache—Sprache als Organismus," paper presented to the conference *Organismus und Technik*, Munich, 1975.

ISBN 0-201-03438-7/0-201-03439-5pbk.

Evolution: Self-Realization through Self-Transcendence

Erich Jantsch

1. PROCESS THINKING

Western science has so far been primarily interested in "point" observations (logical positivism) or, to the extent that it deals with systems at all, in structure (Kant and modern structuralism). Human systems at the physical, social, and cultural level are still mainly described in structural terms—in terms of unambiguous relations between entities which appear as solid, be they houses and roads, or members and subgroups of social organizations and institutions, or values and ideas. Even problems appear as entities which relate to other problems and which can be "solved," just as something solid may be dissolved in a liquid. It is significant that Ozbekhan's (1976) profoundly dynamic notion of a "problématique"—the problematic nature of an evolving situation which becomes manifest in ever-changing elusive aspects which may become transitorily recognized as "problems"—is so frequently misunderstood as an agglomeration of distinct interrelated problems (see, e.g., Fontela and Gabus, 1974).

In this respect, as well as in many others, human systems are frequently described in terms of a paradigm which is borrowed from mechanistic engineering systems. No wonder, then, that in the spirit of this paradigm the preservation of structure is of foremost concern, *equilibrium* appears as a basic condition for achieving it, and *homeostasis* as the cybernetic mechanism to maintain equilibrium. *Negative*, or deviation-reducing, *feedback* is emphasized over *positive*, or deviation-amplifying, *feedback*, and the latter is usually associated with "runaway" systems.

In a structural view, dynamic phenomena are described as processes determined by structure—comparable to a liquid flowing in a fixed piping structure, or a combustion process in an engine block. In this framework, only two basic possibilities can be dealt with, steady-state flow and change in scale (diminution or growth). As has been rightly

ISBN 0-201-03438-7/0-201-03439-5pbk.

Erich Jantsch and Conrad H. Waddington (eds.), *Evolution and Consciousness: Human Systems in Transition.*

observed, growth in scale within a fixed structure runs into the self-evident "limits to growth" (Meadows et al., 1972; Mesarović and Pestel, 1974). If changes in structure are taken into account at all in a structure-oriented view, this is usually described in terms of the cybernetic concept of "ultrastability": the system adapts stepwise via different structural states to an ultimate equilibrium; its evolution is foreclosed by the equilibrium *telos*.

Planning, in the framework of such an engineering system paradigm, focuses either on the preservation (stabilization) of a given structural state by creating and maintaining static system boundaries, or on the definition of a new system state and the appropriate measures to engineer a transition to it—a procedure which tends to put structure logically *before* values and goals, which are generated in a deterministic way by the chosen terminal state. Much of the current effort of normative forecasting and planning is practiced in this spirit of "utopian engineering," as Dunn (1971) terms it. It presupposes that it is unambiguously clear what a "good" future system state would consist of. A structure-oriented approach deals with a social system as if it were a machine to be engineered and controlled from outside; it takes an inherently elitist attitude.

A deeper understanding of natural (contrasted to engineering) systems reveals positive feedback as one of the intrinsic characteristics by which many natural systems—from atoms to galaxies, cells to organisms, social systems to whole populations, single concepts to cognitive systems and whole languages—manage to live and evolve. Laszlo (1972) has attempted to describe all kinds of natural systems by the same four categories of systemic properties: systemic state property (wholeness and order); system cybernetics I (self-adaptation, negative feedback); system cybernetics II (self-organization, positive feedback); and holon property (intrasystemic and intersystemic hierarchy). An important part of natural systems may be characterized as self-organizing (or, perhaps more aptly, self-realizing and self-balancing) systems. In a very general sense, we may call them living systems. From a certain angle of view, they may be described as open learning systems, exchanging energy with their environment.[1] In a more general way, they may be described as self-transcendent systems (See Chapter 2 by Pankow).

In Chapter 5, Prigogine shows for the physical domain that partially open systems in a state of sufficient *nonequilibrium* (so-called dissipative structures) try to maintain their capability for energy exchange with the environment by switching to a new dynamic regime whenever

[1] The concept of energy is used here in a very wide sense, including nonphysical equivalents of energy, such as human motivation, emotions, and ideas, as well as information and complexity.

ISBN 0-201-03438-7/0-201-03439-5pbk.

entropy production becomes stifled in the old regime. This is the principle of "order through fluctuation," which reverses some of the dynamic characteristics holding for closed systems and systems near equilibrium. In particular, order may increase (whereas in closed or near-equilibrium systems, according to the second law of thermodynamics, it can only change toward greater disorder), and the response to fluctuations (which may themselves be random) is the less random the more degrees of freedom the system has. By analogy, dissipative structures may also be recognized at the social level (see Chapter 5) and at the cultural level (Jantsch, 1975).

In a nonequilibrium world of self-realizing, self-balancing systems, process and structure become complementary aspects of the same overall order of process, or *evolution*. As interacting processes define temporary structures—comparable to standing-wave patterns in physics—so structures define new processes, which in turn give rise to new temporary structures. Where process carries the momentum of energy unfoldment, structure permits the focusing and acting out of energy. Only a macro view is capable of providing a perspective of *history*, or evolution of space-time structures; our current microscopic paradigms (e.g., quantum mechanics) do not deal with space-time coincidences.

In the human domain, some of the oldest knowledge systems of mankind, such as Hermetic philosophy, Buddhism, and Taoism were cast in process terms; in their pure form, they did not incorporate structural notions, such as God, essence, or self. But the basic complementarity of process and structure has apparently never been grasped as precisely as now seems possible in the framework of an emerging dynamic general system theory. It is becoming one of the basic paradigms of life in the widest possible sense. Subjects for study which now move into focus include the temporary self-bounding of natural processes (their inherent morphogenetic or systems potential);[2] the complementarity between space-time structure and function (in particular in the interaction between system and environment); conditions for temporary structural stability at discrete levels of complexity; conditions for the penetration or dampening out of fluctuations; time as an evolving system property (not as an absolute); and so forth.

[2] A most important example may be recognized in the study of optimal design of sociotechnological systems and institutions. If such systems are driven too far in size, complexity, and scope, they tend to deteriorate by every measure, even in terms of simple cost-benefit relations; e.g., traveling by car in traffic-congested cities may cost more time than walking. Illich (1975) has termed this phenomenon "specific counterproductivity" and explains it by the increasing dominance of *heteronomous* ways of production (supply of ready-made, standardized goods and services) which tend to inhibit the self-bounding nature of *autonomous* ways of production (learning, doing it yourself).

ISBN 0-201-03438-7/0-201-03439-5pbk.

The concept of dissipative structures also implies that the environment is necessary to maintain the dissipative structure; unlike equilibrium structures, dissipative structures cannot exist independently. But then, evolution necessarily includes the environment of any evolving system under consideration. Biological evolution already may be understood as "the generation of organisms with nonrandom behavioral responses to the environment. The whole thrust has been toward improving the correlation between the environment and the behavior of organisms" (Dunn, 1971). The history of ecosystems and the whole biosystem may be viewed as an aspect of multifold phylogenesis. In the human world, there is a very strong two-way interaction in the much faster processes of technological intervention in the environment, and the conditioning of man by his technologically changed living space. Such a large-scale feedback interaction in the human world now seems to demand a shift from the one-way causal concept of energy-pushing technology (changing the environment) to the mutual-causal concept of a *cybernetic technology*—recycling, tapping natural energy flows (hydropower, solar and tidal energy) instead of liberating energy stored in fossil or nuclear fuel, biological production processes beyond agriculture, and so forth.

But the correlation between human behavior and the physical environment corresponds to just one out of many levels of human existence. Similar two-way correlations prevail at the level of the social, the cultural, and generally also the spiritual environment. Human evolution is both unfolding of an inherent dynamic potential, and correlation with many levels of dynamic environment which, in their totality, fall together with universal evolution.

2. MODES OF LEARNING

This interaction between a system "phenotype" and its environment at the physical, social, and spiritual levels may be described in terms of learning processes in a general sense. The prerequisite for learning is a certain plasticity of the system which allows it to have history. In evolving systems, or systems with history (e.g., all dissipative structures), each system state depends on the past development of all subsystems. There exists therefore a "system memory" in terms of the significance of temporal anisotropy in the dynamic structure, a very different idea from a "data bank" type of memory. Each system with history "remembers" the total development of itself as a whole—and is thus also capable of the *re-ligio* to its own origin.[3]

[3] Such a "system memory" is also implied by recent concepts (Fischer, 1971; Horn et al., 1973) which emphasize that experience is state-bound, and therefore

ISBN 0-201-03438-7/0-201-03439-5pbk.

Different basic kinds of learning processes may be distinguished. One way of distinguishing them would be by physiological features (e.g., membranes, neurons, neuron networks, cortex in evolutionary order). However, it appears of advantage here to choose a classification which is less ambiguous and overlapping with respect to consciousness. Following Laszlo's (1972) categorization of natural systems by the type of consciousness they bring into play, ontogenetic learning processes may then be characterized by the way in which system consciousness unfolds in them:

- *Virtual learning* is characteristic of nonreflective consciousness (which Laszlo terms "subjectivity") and physical development. For example, an atom in transition to another system state "lives" simultaneously many virtual processes within the margins set by quantum-mechanical indeterminacy; or a fluctuating dissipative structure "tests" in this way its possibilities for shifting to a new dynamic regime, for which it has generally at least two options.
- *Functional learning*, characteristic of reflective consciousness or simple perception, may be graphically depicted by the feedback interaction between consciousness and environment which appear here in a binary link. This kind of interaction is found in biological and primitive social processes. As Fischer (1975a) points out, "the process of translation from messenger RNA to polypeptide chains carries only one-dimensional information in the sequence of the amino acids. It is the interaction of this one-dimensional text with its 'physiological environment'—a hypothesis testing situation—which determines three-dimensional texture." When a single-cell organism moves in an environment, it does so by "testing" the immediate environment in numerous short steps in all directions before making a big step in one selected direction, "vindicating" the validity of the choice only after venturing out so boldly. I call this type of learning functional because what is tested here are primarily metabolic (energy exchange) functions in terms of relations with the environment. It is highly significant that recent findings in biology seem to suggest that the genetic program is applied not so much to the generation of behavioral patterns as to the fulfillment of certain vital functions in a situation-contingent way.[4]
- *Conscious learning*, characteristic of self-reflective consciousness or apperception, may be viewed as the multiple feedback interactions in the ternary system formed by consciousness, the environment, and a memory or storage system which may be termed the "appreciated world" and which, of course, is itself part of consciousness, but is set aside here for reasons of convenience (for an elaboration of this model, see Jantsch, 1975). Conscious learning is the normal mode of

can be "relived" when the corresponding psychological/physiological state is induced. Experiencing a work of art seems to have much to do with "tuning in" to the human state in which the work was conceived.

[4] For example, in an experiment in which the function of the ventilation system in a termite mound was impeded by placing the whole structure under a plastic tent, it took the termites less than two days to come up with a totally new design which fulfilled the function of temperature control again in a satisfactory way (Frisch, 1974, quoted in Waddington, 1974).

ISBN 0-201-03438-7/0-201-03439-5pbk.

learning and of becoming creative in the human social realm. In the appreciated world we test and develop ideas and plans which then allow us to control the focusing and direction of physical and emotional energy in transforming the environment.

- *Superconscious learning*, characteristic of a more complex kind of self-reflective consciousness which mirrors itself in a "surface" consciousness as well as in a multilevel superconsciousness or "depth" consciousness.[5] It may be depicted by the feedback relations within a quaternary system, or initially perhaps by the same ternary system as conscious learning, but with a distinction made between "outer" and "inner" ways of learning. Superconscious learning provides a sense of direction for cultural and mankind processes by "illuminating" the process from the far end in terms of guiding images (see Chapter 11 by Markley and 12 by Jantsch).

These learning modes characterize *levels of macroscopic coordination* of interactive processes. Matter itself has come to be viewed as an aspect of such a multilevel process coordination—not elusive structures, but the *relations* between coordinative levels move into focus now. In the energy band and under conditions conducive to the types of macroorganization found in biological life, the laws of physics become accentuated in particular ways by a hierarchy of coordinative levels ranging from dissipative structures at the molecular level and within cells to bio-organisms, societies, ecosystems and beyond. Even "mind" may now perhaps be understood as higher-level coordination of the same processes which, at other levels, appear as "matter"; thus, a duality vanishes that has long haunted Western thought. In such a perspective, evolution relates to *change in organization* rather than substance, and the evolution of life to the development, at various levels, of appropriate modes of knowledge generation, transfer and utilization in a semantic context. Eigen and Winkler (1975, p. 120) suggest that the same basic mechanism of self-transcendence operates at all levels of life: "Dissipative processes regulate and synchronize the retrieval of information stored in conservative structures and guarantee its functional effectiveness." In other words, *in-formation*, or phylogenetic history frozen into syntacti-

[5] Superconsciousness is understood here as that transpersonal aspect of total consciousness through which we share in group, cultural, and mankind processes—or even in universal evolution. C. G. Jung's archetype-carrying "collective unconscious," Assagioli's "Higher Self" (Assagioli, 1971), the *alayavijnana* (store-consciousness) of Buddhism and the *atman* of Vedanta are related notions, as in Fischer's (1975b) multilevel concept of "self"-awareness through increased subcortical arousal. The feedback link between consciousness and superconsciousness gives rise to "inner experiential learning" or "tuning-in" to the dynamics of metasystems transcending man and his immediate environment. It may be enhanced by various techniques, mostly developed in connection with Eastern philosophies (for an overview, see Jantsch, 1975, Chapter 9; and Naranjo, 1972).

ISBN 0-201-03438-7/0-201-03439-5pbk.

cally ordered form (such as a genetic code, a social law, or a work of art) is "brought to life," or placed in a time perspective, by *in-struction* from a higher level of coordination (bringing into play semantics, or the functional relations of a space-time structure to its environment); the result is *in-tuition*, or learning from within by *re-ligio* (connecting through a space-time continuum to one's own origin)—ontogeny recapitulates phylogeny in the life of art or ideas no less than in biology. The deterministic factor in self-organization (or "canalization" in Waddington's terms) is itself the product of the history of that same self-organization and thus of the interaction of stochastic and deterministic factor, chance and necessity. (See also Chapter 2 by Pankow for discussions of the information-instruction hierarchy and of *re-ligio*.)

In a *phylogenetic* perspective, virtual learning may be correlated with the development of duplicative modes ranging from prebiotic self-reproducing hypercycles (Eigen, 1971) to cell division. Evolution works slowly by mutation and error-amplification and selection acts on populations of little varied genetic types. For early prebiotic stages, Eigen and Winkler (1975, p. 259 ff.) even assume hyperbolic growth of one type and "all-or-none" selection. Functional learning accelerates evolution by developing sexual reproduction (mixing within gene pools) and establishing cybernetic links with the environment (adding epigenetic information). Variety increases greatly as well as the capabilities for symbiosis and adaptation. Selection here acts on intra- and interspecific sociobiological systems (societies and ecosystems). Conscious learning develops language, from simpler forms of animal expression to the semantic richness of human language, thereby further accelerating evolution beyond any genetic pace of change and enhancing systems coherence and resilience. The development of ideas as whole gestalt systems greatly increases the scale and effectiveness of action and thus also phenotypal flexibility. Selection acts on sociocultural systems (epistemologies together with their supportive social institutions). Superconscious learning, finally, develops modes of sharing in "Mind-at-Large" (Huxley, 1954); the subject of selection is humanity.

Results from serial LSD therapy (Grof, 1975) seem to suggest that the four learning modes may also be understood in terms of an *ontogenetic model of human consciousness*, including, in particular, a perinatal phase as a complete life cycle before life on earth (see Figure 3.1).

Grof's scheme of four perinatal matrices of experience (represented in Figure 3.1 as fields F-1 to F-4) also seems to bear out the archetypal quality of the first four numbers as ordering principles of a psychophysical one-continuum (which becomes manifest in complementary physical and psychic aspects), known by ancient cultures and rediscovered in our time by C. G. Jung and Marie-Louise von Franz. According

ISBN 0-201-03438-7/0-201-03439-5pbk.

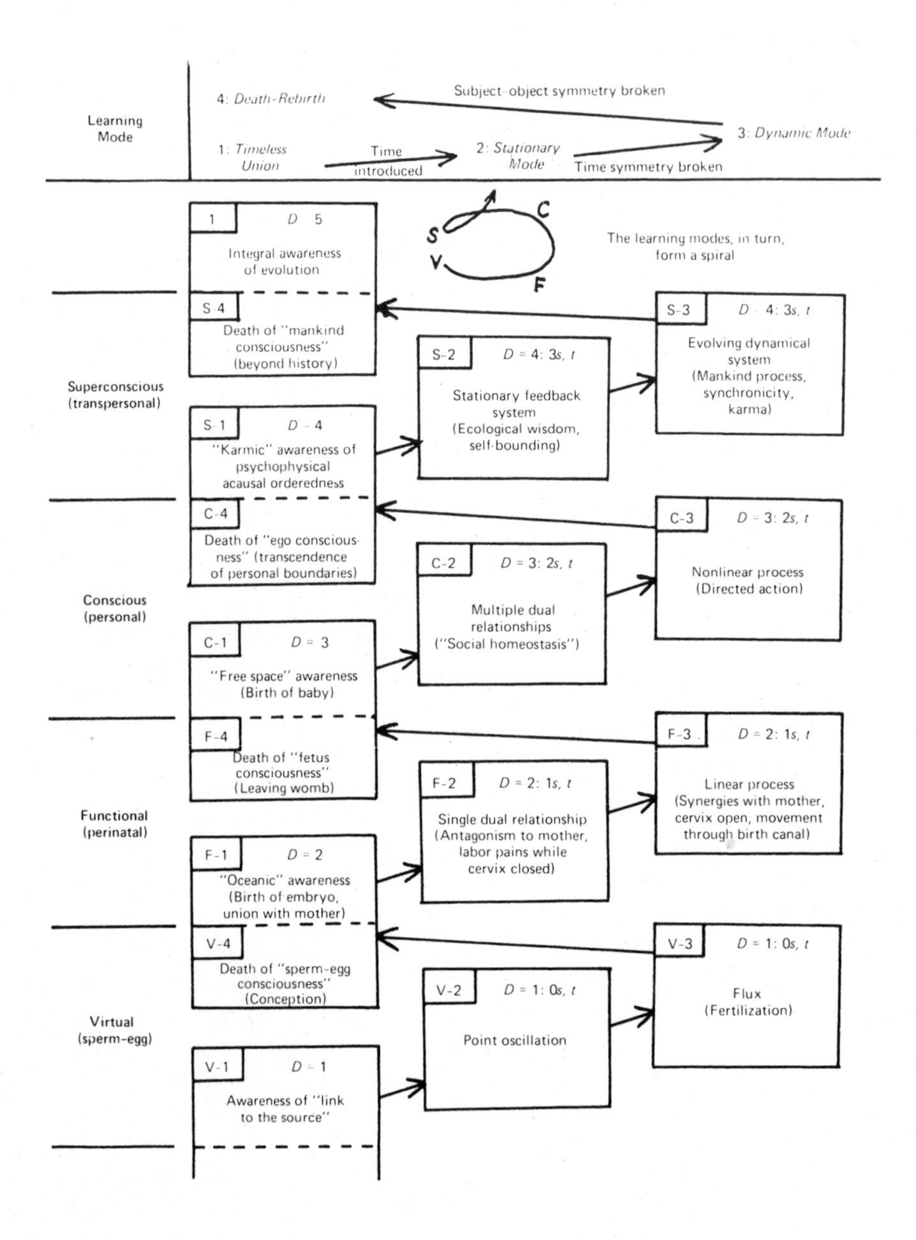

FIGURE 3.1. An ontogenetic model of human consciousness—the double learning spiral. (D number of dimensions, s spatial dimensions, t time).

ISBN 0-201-03438-7/0-201-03439-5pbk.

to Franz (1974, p. 74), the time-bound qualities of the first four numbers are as follows: "One comprises wholeness, two divides, repeats, and engenders symmetries, three centers the symmetries and initiates linear succession, four acts as a stabilizer by turning back to the one as well as bringing forth observables by creating boundaries, and so on." The transitions between these four basic qualities (which constitute a system rather than a logical sequence) symbolize how a gestalt system maintains its nature in the presence of many temptations to become formalized. The first step from one to two constitutes the "original sin" of formal division which "separates the two sides of complementarities and treats them as identities" (Pankow, p. 35). Spencer Brown (1969) has shown that this step establishes the identity of organization of the outer and the inner world; this is the beginning of formal description, not of gestalt perception. In a fascinating thought experiment, by means of a new "calculus of indications," he delineates the generation of a universe of forms following the severance of space (e.g., when an observer reflects on the world and thus breaks the original unity): "By tracing the way we represent such a severance, we can begin to reconstruct, with an accuracy and coverage that appear almost uncanny, the basic forms underlying linguistic, mathematical, physical, and biological science, and can begin to see how the familiar laws of our own experience follow inexorably from the original act of severance. The act is itself already remembered, even if unconsciously, as our first attempt to distinguish different things in a world where, in the first place, the boundaries can be drawn anywhere we please. At this stage the universe cannot be distinguished from how we act upon it, and the world may seem like shifting sand beneath our feet. Although all forms, and thus all universes, are possible, and any particular form is mutable, it becomes evident that the laws relating such forms are the same in any universe. [p. v]." Therein lies a formal justification for the ultimate complementarity of the search without (in the physical world) and the search within (in our own experience), "for what we approach, in either case, from one side or the other, is the common boundary between them [p. xix]." The same idea has been expressed in Vedanta by the identity of *brahman* and *atman* (see also the section on "Complementarity" below).

Of particular significance in the unfolding of archetypal qualities, or basic human experience, is the dynamic quality of the number three (Franz, 1974, p. 106): "Three signifies a unity which dynamically engenders self-expanding . . . irreversible processes in matter and in our consciousness (e.g., discursive thought)." But the step from three to four is the most important one, a "powerful retrograde connection to the primal one [Franz, 1974, p. 129]." In it, gestalt systems reestablish

ISBN 0-201-03438-7/0-201-03439-5pbk.

themselves as individuals which have their complementarities within themselves and which are "identical neither with each other nor with themselves; nor . . . strange to each other" (Pankow, p. 35).

The same scheme underlying perinatal life also applies to human life on earth, characterized primarily by conscious learning. The Tree of Life of the Cabala structures a life cycle which spirals up in three complete turns before it returns, in the fourth phase, through its own center (Purce, 1974, Figures 41 and 42). According to Franz (1974, p. 124ff.), Jung linked specific human attitudes to the first four numbers in the following way: "At the level of one, man still naively participates in his surroundings in a state of uncritical unconsciousness, submitting to things as they are. At the level of two, on the other hand, a dualistic world- and God-image gives rise to tension, doubt, and criticism of God, life, nature, and oneself. The condition of three by comparison denotes insight, the rise of consciousness, and the rediscovery of unity at a higher level; in a word, gnosis and knowledge." But, in the ternary model, thinking is "flat, intellectual . . . an element shining forth as 'timeless structure' in the unconscious has been reconstructed through discursive thought processes, and, in this process, became temporally conditioned." It is only the step from three to four which conveys a sense of wholeness and a perspective of inner and outer reality which is capable of guiding our thinking within a wider existential context. In a similar way, I have distinguished (Jantsch, 1975) between rational, mythological, and evolutionary levels of inquiry, which correspond to C-2, C-3, and C-4/S-1 in Figure 3.1.

Figure 3.1 attempts to sketch in a highly simplified way the ontogenetic development of human consciousness as a spiral of (at least) four death–rebirth cycles, from sperm–egg through perinatal and personal to transpersonal consciousness. Each of the four life cycles is in turn ordered by the development of qualities corresponding to the same fourfold scheme outlined above for the perinatal and personal life cycles. The fourth phase always falls together with the first phase of the next stage. Ontogeny seems to recapitulate here the development of learning modes, or evolutionary mechanisms of correlating individuals with their environment, which are quite universal.

At the same time, this scheme seems to depict the stepwise gain of new dimensions for relating to the world, so that the experience of wholeness in phase 1/4 evolves from that of a one-dimensional link to the source through two-dimensional "oceanic" awareness (Grof, 1975, p. 106f.) and three-dimensional "free space" awareness (Grof, 1975, p. 139) to four-dimensional karmic awareness,[6] or awareness of the

[6] Karmic bonds, which play an important role in Eastern cosmologies, may be described as spatially and temporally disjointed feedback relations. For example, what a person sows in a particular situation, he/she may reap in an apparently to-

ISBN 0-201-03438-7/0-201-03439-5pbk.

Jungian "acausal orderedness" (Grof, 1975, pp. 174f.), and further on to an awareness of evolution which includes its beginnings and ends. In phases 2 and 3 of each stage, this D-dimensional awareness becomes structured in a time-bound way in terms of (D - 1) spatial dimensions and one temporal dimension. For example, a two-dimensional "oceanic" awareness becomes structured into a single dual relationship in the stationary state of phase 2 (good or bad relationship of fetus with mother, and in particular labor pains with the ensuing fetus–mother antagonism), and into a linear process in the dynamic phase 3 (the now open cervix provides one, and only one, "way out" and fetus–mother synergism initiates an irreversible process of movement through the birth canal). In this way, phase 2 experiences develop from point oscillation through single and multiple dual relationships to the experience of a complex stationary feedback system; and phase 3 experiences develop from flux through (spatially) linear and nonlinear processes to the experience of integrally evolving systems.

It is interesting to note here that this scheme also generalizes the symmetry-breaking nature of the steps taken to new levels of inquiry which have been pointed out in the domain of physics (Prigogine, 1973) as well as in social inquiry (Jantsch, 1975). In the transition from phase 2 to phase 3, the time symmetry is broken; irreversibility is introduced. In the transition from phase 3 to phase 4, the symmetry between subject and object is broken; a nondualistic experience marks the gain of a new dimension.

Of course, the development of human consciousness in the individual does not go through this spiral of stages and phases in a single, irreversible process. Already in perinatal life various disturbances may introduce fluctuations between F-1 and F-2. In our personal life, we switch frequently between C-2 and C-3, but we can also experience repeatedly the death–rebirth step between ego consciousness and transpersonal consciousness. In fact, whereas virtual and functional learning are characteristic of particular physical stages of human life, in our life as a person all modes of learning are open to us, in particular, a combination of conscious and superconscious learning, or discursive thought against a background of understanding the wholeness in which human life is embedded.

This combination of cortical and subcortical activity has been described by Fischer (1971, 1972, 1975b) as communication between the

tally unrelated situation, possibly separated from the first one by several reincarnations. Karma works like an accounting scheme for evolution in its totality, an idea expressed in the image of the akashic records in which all "credits" and "debts" are written down. In modern terms, we may perhaps view karma as psychophysical *chreods*, a term suggested by Waddington (1975, p. 22) and referring to "a canalized trajectory which acts as an attractor for nearby trajectories."

ISBN 0-201-03438-7/0-201-03439-5pbk.

"Self" (subcortical—whole organism) which sees and knows, and the "I" (cortical—brain) which interprets that which is seen and known by putting it into the context of physical space-time. Such a communication underlies all art, science, and religion. The way to the "Self," or to superconscious learning, is twofold: through increased ergotropic (ecstatic) arousal, characteristic of creativity and of Western forms of mystical experience; and through increased trophotropic (meditative) arousal, characteristic of dreaming and of typical Eastern forms of mystical experience. Generally, the notions of the "I" and the "Self" are correlated with left-hemispheric (analytical, sequential) and right-hemispheric (holistic, cognitive-mapping) functional emphasis of brain activity, respectively (Fischer and Rhead, 1974). To some extent, the left-hemispheric rational/analytical mode appears to be the result of cultural learning—like a special form of expression learned by the basic gestalt system and displacing "the non-sequential functions of cognitive-mapping . . . from the left speech-motor areas" (Fischer and Rhead, 1974). Gestalt perception, in this view, is then not restricted to the brain, but a function of the organism as a whole. Mind starts looking like an ecosystem. Fischer (1975b, and personal communication) even proposes the interesting hypothesis that hemispheric functional emphasis has shifted repeatedly in the history of Western civilization, with right-hemispheric (matrist) ages marked by the start of the thirteenth century (Renaissance), the Elizabethan Age, and the second half of the twentieth century. The changeover periods of balanced interhemispheric integration give rise to bursts of creative energy (e.g., in the first decades of the twentieth century). However, Western science is still officially restricted to the "logical" rational/analytical mode and even disqualifies any version of "intuitive" gestalt perception, although the latter has played such a significant role in other cultures and, one may add, in the significant breakthroughs and novel conceptions of that same Western science by which it is now condemned. In this situation, as Fischer and Rhead (1974) point out, "the IQ has come to be essentially a ratio of left to right hemispheric capacity" and thus a measure of imbalance. The consequences for a reorientation of educational policy appear to be tremendous.

It is this unique combination of two learning modes which gives social systems and cultures in the human domain their flexibility to evolve. It is here that our hope lies for "tuning in" to the structure and dynamics of the metasystems of human life, and thus for acting not just with efficiency, but also with wisdom. A four-dimensional awareness of what Jung has termed the timeless "acausal orderedness," when translated into time-bound experience (insight), leads in the stationary case S-2 to ecological wisdom, or an "instinctive" recognition of the complex patterns of interaction and the self-bounding character of natural

ISBN 0-201-03438-7/0-201-03439-5pbk.

processes. In the dynamic case S-3, it leads to a beyond-rational grasp of evolving systems, in particular, of groups, cultures, and mankind as an integral process. Such a grasp also often includes some insight into karmic structures or "chreods" in transpersonal processes, and into the nature of the coincidence of physical and psychic phenomena which Jung has termed *synchronicity*.[7]

There is one more difficulty to resolve: In LSD therapy (Grof, 1975), the deepest stratum of the human unconscious—the one which is generally revealed last—contains both experiences of a prehuman and subpersonal kind (such as identifying with a sperm cell, or with parts of the body) and experiences of a transpersonal kind (such as "living" the phylogenetic history of man, identifying with group or mankind processes, or insight into the mechanisms of the evolution). But in Figure 3.1, these experiences would belong partly to virtual learning and partly to superconscious learning. Obviously, the linear scheme of learning modes, convenient for purposes of representation, is misleading; there is another retrograde from 4 to 1.

Just as the evolution of particular learning modes may be drawn as a spiral, with the step to phase 4 always implying a linking backward to the primal one, superconscious learning—the quaternary model of consciousness, the four-dimensional awareness—links up with virtual learning, the one-dimensional instinctual life. The learning spiral is a *double spiral.*[8] Going forward becomes at the same time *re-ligio*, the linking backward to the origin. It is this deep thought which Heinrich von Kleist has formulated in his essay "On the Puppet Theatre" when he observes that in the human domain gracefulness returns when cognition has gone through "an infinite," just as the image in a concave mirror, after having vanished into infinity, suddenly returns close before our eyes. Thus, says Kleist, gracefulness is most purely expressed through a

[7] Synchronicity is the time-bound manifestation of a psychophysical acausal orderedness: "Just as causality describes the sequence of events, so synchronicity . . . deals with the coincidence of events . . . . How does it happen that A′, B′, C′, D′, etc. appear all in the same moment and in the same place? It happens in the first place because the physical events A′ and B′ are of the same quality as the psychic events C′ and D′, and further because all are the exponents of one and the same momentary situation [Jung, 1949, pp. xxiv–xxv]."

[8] After finishing the manuscript of this chapter, I became aware of a strikingly similar concept expressed by Fischer (1966, pp. 372ff.). He views life on earth as "a single event, exponentially receding and proceeding in time." This unity of life is symbolized by an exponential (logarithmic) spiral of time on which are ordered "adaptive events of increasingly efficient utilization of energy as well as increasingly rapid time rate of change." These adaptive events are grouped into four categories (in the inward spiraling sense evolution, learning, perception, and hallucination/dreaming) which, in spite of the different use of words, turn out to be the precise equivalents of the four learning modes discussed here. The first and fourth class of adaptive events also touch in Fischer's spiral.

ISBN 0-201-03438-7/0-201-03439-5pbk.

body which holds no self-reflective consciousness at all—or infinite self-reflective consciousness. "Consequently, I said a bit distracted, we would have to eat again from the tree of knowledge in order to return to the state of innocence? Indeed, he answered, this is the last chapter of the history of the world."

The autocatalytic principles at work in the different modes of learning/creativity may also be characterized by the following interlinking hierarchy, given here in ascending order:

- Virtual learning: movement to change movement (response);
- Functional learning: response to change response (behavior);
- Conscious learning: behavior to change behavior (conscious action);
- Superconscious learning: consciousness to change consciousness (superconscious self-regulation of cultural and mankind processes).

Two highly significant characteristics apply to all four modes of learning: First, learning is always *open-ended*, never deterministic; it is geared to a process of open-ended experimentation, comparable to the strategic exploration of available options with subsequent vindication of the choice made—in other words, it is *learning by doing* in a partially informed way, guided by higher modes of learning; thus, it is an aspect of self-transcendence (see Chapter 2 by Pankow). Second, what is learned is not places or states or relations, but a *sense of direction;* thus, evolutionary learning inherently refers primarily to process, not to structure. These two characteristics may be combined in the notion of *evolutionary experimentation.*[9]

Time is involved here in a particular way which tends to facilitate evolutionary experimentation: Ascending the learning hierarchy, strategic "forecasting" of change becomes increasingly long range—from near-infinitesimal fractions of a second in virtual learning all the way to aeons in superconscious learning (see Chapter 12 by Jantsch). At the same time, evolution of the system "genotype" runs ever faster as we climb the hierarchy, from billions of years for chemical and millions of years for biological evolution to hundreds of years for social evolution in the prescientific period and mere decades for cultural evolution in our present period in which superconscious learning assumes increasing importance. But this is a fortunate combination: the faster we go, the farther ahead we have to look for signs to guide our movement. It points also to the increasing flexibility of the "phenotype" or individ-

[9] Sackman (1967) and Dunn (1971) employ this term at specific levels of conscious and superconscious learning only, because they restrict notions such as "learning" and "experimentation" to the human world. I believe that a more profound understanding of evolution results from a multilevel view in which human consciousness forms but one level of typical action and interaction.

ISBN 0-201-03438-7/0-201-03439-5pbk.

ual system as we proceed from the physical through the biological and social to the cultural realm, confirming this increase in flexibility perhaps as one of the principal aims of evolution (see also Chapter 1 by Waddington).

3. HIERARCHICAL DESCRIPTION: MULTILEVEL LEARNING

Hierarchical descriptions of *structure* usually express the ideal of *equilibrium* both at a given structural level and in the relations between levels. If structures at a given level try to maintain homeostasis with respect to reference values (or goals) defined "from above," we have the crude organismic system concept; the only level which, in this concept, is free to venture out and "invent" changes of reference values, is the top level—an organism, the environment of an organism, or any metasystem depending on how the system boundaries are defined in an *ad hoc* way. In the human world, this concept translates into dictatorship. If structures at a given level try to maintain homeostasis with respect to reference values (or goals) defined "from below," we have the multilevel, multigoal (multiechelon) system concept in the version for which Mesarović et al. (1970) have developed a mathematical theory; here, the only level free to "invent" goals is the bottom level, whereas all higher levels and the system as a whole derive their goals from a stepwise coordination of lower-level goals. A special case may be found in a system which tries to equilibrate its levels in an ascending direction, but denies its bottom level any further deviation from internal equilibrium.[10]

A description in terms of structure has to assume that there are clearly separable goals, or consistent sets of goals. A "top-down" approach further assumes that the supreme goal held by the top level may be broken up neatly into subsystem goals which contribute to it. A "bottom-up" approach assumes coordinability of subsystem goals into a single-goal concept at the next higher level. Generally, individual members of the system will then be considered to belong to one subsystem only at each level. But this is all far from social reality in the human domain where multiple membership in institutions is the rule and where balancing the multiple demands arising from such a system

[10] Contemporary writings in social science abound with applications of a scheme which may be called "filling the needs/values hierarchy from the bottom up"—especially in terms of Maslow's hierarchy, which ascends from physical needs (physiological needs or survival, and safety) through social needs (belonging and esteem) to spiritual needs (self-fulfillment).

ISBN 0-201-03438-7/0-201-03439-5pbk.

of relations even constitutes "the basic art of being human" (Vickers, 1973). Values, and goals corresponding to their realization, may be more aptly thought of as emerging from a field between opposites which cannot in a lasting way become clearly resolved at either extreme (Jantsch, 1975).

In contrast, a *learning hierarchy* focuses on feedback processes between the levels which are then also understood as forming a hierarchy of intra- and intersystemic processes. Such a *process hierarchy* in the human world may then be viewed, for example, as sketched in Table 3.1, open in both directions, upward as well as downward. Elsewhere (Jantsch, 1975) I have described the "waves of organization" which successively enter the human world, orchestrating it ever more fully. Out of the struggle of the individual for physical survival grew social systems which, in turn, made it possible to organize physical relations in a conscious approach, involving human design. Out of the struggle within and between social systems grew cultures which, in turn, provided an ethical roof beneath which to organize social relations in a consciously designed way. Out of the struggle within and between cultures—a struggle of images which man holds of himself, of life styles and world views—grows in our time a feeling for the wholeness of the global mankind process. Internal, coordinative factors and external, competitive (Darwinian) factors in evolution which have been distinguished at the level of the individual organism (Whyte, 1965) become complementary aspects of the evolutionary process, depending on the system level from which it is regarded. The external, interindividual struggle becomes the internal, intrasystemic factor in the emergence of social systems, and so forth from level to level.

Ecological, social, and cultural organization may be viewed as being initiated in a subconscious way—in our time, cultural organization (value dynamics) is still in this phase, then passing through a predominantly externalized phase of competition, and becoming internalized again by conscious coordination, or human design.

This interaction of staggered phases in multilevel processes of organization may now also be understood in terms of the learning hierarchy: The hominid world developed first primarily in the *functional* learning mode characteristic of the whole bioorganismic world, of *phylogenesis* in the biological meaning, which is closely linked to biological genetic communication and evolution. Phylogenesis is not just a matter of bioorganismic organization, but also of ecological organization—as each phylum evolves with its operational and potential environment, ecological organization comes into play. Through functional learning, social systems (including animal societies) may form.

Conscious learning seems to be the primary vehicle for epigenetic development and an important factor in sociobiology. Recent research

ISBN 0-201-03438-7/0-201-03439-5pbk.

ISBN 0-201-03438-7/0-201-03439-5pbk.

TABLE 3.1 Multilevel Learning in the Human World

Level of Learning	*Level of Organization*	*Complementary Perspective*	*Systems Level*	*Evolutionary Process Level*
⋮	⋮	⋮	⋮	⋮
			Mankind	Anthropogenesis
		internal to		
	Cultural		g h	
		external to		
Superconscious (Self-Regulation)			Cultures (Noosphere)	Noogenesis
		internal to		
	Social		e c d f	
		external to		
Conscious (Action)			Social Systems (Sociosphere)	Sociogenesis
		internal to		
	Ecological		a b	
		external to		
Functional (Behavior)			Individuals	Phylogenesis
		internal to		
	Bioorganismic			
Virtual (Response)				
	⋮	⋮	⋮	⋮

seems to establish that at least higher animals form mental models through which they relate to the world.[11] This conscious and creative interaction of the individual with its environment seems to explain the increase of individualism in the evolution of social animals, which is certainly not the negative factor as which it is presented by Wilson (1975). In the human world, conscious learning "reformed" ecological organization and made it a matter of conscious and elaborate design (agriculture, etc.). It is central to *sociogenesis*, which is increasingly becoming a matter of human design, too, and brings culture into play. In this advanced human stage, conscious learning is closely linked to the semantic richness of language as the proper mode of sociogenetic communication and evolution, transmitting a vastly expanded heritage of evolution, besides the genetically transmitted heritage. As Waddington (1975, p. 278) argues, ethical beliefs necessarily develop with this new evolutionary mechanism. The gradual shift of emphasis from linear genetic to intra- and intersystemic epigenetic development with the unfolding of the learning modes implies a trend toward fuller expression of the organism as a whole.

The next learning mode, *superconscious learning*, provides guidance for these ethical beliefs. Superconscious learning, accessible in an experiential way, is already strongly present in the later stages of social organization by way of morality (which is the learning of "inner" constraints) and archetypes, which are a symbolic expression of basic humanness (or norms inherent in the mankind process). Beyond all cultural bias, we "know" in a very profound way what is morally right, without referring to any explicit ethical code. In other words, we carry the coordinative conditions of human system development within ourselves—if we feel "in tune" with evolution, we need no outside reference. Superconscious learning is central to *noogenesis*.[12] In a time of transition and strongly felt value dynamics, it becomes evident to what extent superconscious learning dominates the lives of humans and

[11] I owe this information to Prof. Walter Freeman (University of California, Berkeley), who found in his electroencephalographic (EEG) research on cats, rabbits, and similar species that there is great brain-wave commotion when an animal is forced to pay attention to something it has consistently ignored until then. A new EEG pattern (a new model of the situation and its context) appears fairly quickly and is subsequently adjusted until it satisfies the new set of observations. Prof. Freeman believes that model building must play a role for all animals with locomotion. Life, including mental life, is not just about "fitting" and deduction, but about the generation of new concepts.

[12] The term "noogenesis" is employed here to denote the becoming of man's cultural/mental world. It is also used in a much wider sense by, e.g., Teilhard de Chardin, who meant by it "consciousness of consciousness" of the universe.

ISBN 0-201-03438-7/0-201-03439-5pbk.

human systems—even if it may appear as the play of an irrational and capricious Zeitgeist until we have learned to listen to it sufficiently well to gain a clearer sense of movement and direction. We cannot yet clearly divine the values inherent in mankind's evolution. "Survival of the species" is a trivial equilibrium view. A nonequilibrium view would emphasize the creative, outgoing—and joyful—aspect of life and place mankind learning in perspective as part of superordinate, more universal learning processes. Perhaps higher kinds of superconscious learning will ultimately open up our knowledge of these metadynamics and help us to establish the dynamic mankind system in a meaningful, that is, purposeful, way.

A specific level of learning is geared to the adaptation of the individual to the corresponding level of environment—physical, social, or cultural. In this stage, freedom is mostly expressed through a limited choice of the habitat, or the "exploitive system" (Waddington, 1975, pp. 284f.), in all three domains; the limitation in the choice of the social habitat (e.g., choice of profession or milieu) is, of course, a main theme of contemporary social and political theory.

But a switch to the next higher level in the learning hierarchy opens up the possibility of adapting the environment to the individual, thus completing the two-way correlation between the individual and a specific level of environment. In other words, a switch to a higher mode of learning opens up a new level of *design*. The intricate interaction between different modes of learning, as sketched in Table 3.1, seems to ensure that the correlation always becomes complete in this two-way sense and that it is continued at higher levels of environment:[13] As functional learning adapts man to his physical environment, conscious learning adapts the physical environment to man (mainly by means of technology). As conscious learning adapts man to his social environment, superconscious learning of the type we experience at present in a widely shared way (through guiding images, values, and ethics) adapts the social environment to man; and as superconscious learning adapts man to his cultural environment, perhaps a higher mode of superconscious learning will make the cultural environment a matter of human

[13] Bateson (1972, pp. 361f.) emphasizes the *centrifugal* direction of biological evolution moving from "adjusters" (e.g., poikilothermic animals, which adjust their body temperature to the environment) through "regulators" (e.g., homoiothermic animals, which keep their body temperature constant) to "extraregulators" (man with his capability of changing the environment significantly). It is interesting to note here that centrifugal evolution is favored by variable stress and thus incorporates a basic principle of nonequilibrium and fluctuation, whereas constant stress would favor a centripetal direction of evolution.

ISBN 0-201-03438-7/0-201-03439-5pbk.

and social design, too. Thus, we may speak of a metaevolution of biogenetic, sociogenetic, and noogenetic mechanisms—of an evolution of modes of evolution, or of learning how to learn better.

The levels of learning turn out to be *levels of openness.* At each new level a new enterprise starts which transcends the order unfolding at the previous level—and usually disturbs it, or as it appears now, furnishes fluctuations encouraging its further evolution. Systems are open to such fluctuations in many dimensions because different modes of learning (of openness) interpenetrate each other in any particular system.[14] Each system level emerges from no less than *three* types of learning in the scheme of Table 3.1—and maybe from more. Sociogenesis, in this perspective, is the result of overlapping functional, conscious, and superconscious modes of learning. Social systems may receive fluctuations from changing values as well as from changing social and physical relations. Therefore, in trying to model sociogenesis—an upcoming great task for social science—it is of the utmost importance first to clarify which aspects may be represented by which kind of learning.

Out of this intermeshing operation of learning loops, which may be called a consequence as well as a description of self-transcendence, arises a sense of direction beyond the scope of a particular system level, and beyond its time horizon. As we are setting out at present to correlate ourselves with our cultural environment through superconscious learning (links e and f in Table 3.1), the higher loop of the same learning process (links g and h) already feeds us an increasingly dense stream of guiding images from the mankind level, and has done so for quite some time in human history (see also Chapters 11 by Markley and 12 by Jantsch).

If successive stages of environment become correlated with man in a two-way process, man himself becomes an expanding concept. Looked at in this way, evolution may then also be understood in terms of an

[14] To what extent such fluctuations represent "chance" is not clear, especially in the domains of conscious and superconscious learning. As Franz (1974, pp. 230f.) remarks, "it is just in such 'chance' occurrences that startling new ideas erupt. It is being more and more firmly established that parapsychological phenomena occur mainly in the surroundings of an individual *whom the unconscious wants to take a step in the development of consciousness,* as, for instance, adolescents who must take the 'leap' into adulthood. Creative personalities who must fulfill a new creative task intended by the unconscious also attract such phenomena. . . . This means that whenever a creative intention is present in the unconscious, parapsychological and particularly synchronistic phenomena, which Jung calls 'acts of creation', may be expected." Of course, what springs to the mind here instead of the concept of an "intentional" unconscious is the amplification of fluctuations in intensely "alive" dissipative structures under certain conditions (see Chapter 5 by Prigogine)—but the results in the human world are often such that they seem to suggest, in addition, a teleological or a psychophysical "field" concept.

ISBN 0-201-03438-7/0-201-03439-5pbk.

expanding multilevel process concept of self-transcendence—or, viewed as a totality, as an overall process of *self-attunement* of an evolution which becomes ever more fully self-reflective, conscious of its own unfolding. As Smuts (1926) had already formulated fifty years ago, the assumption of a single principle, holism, is capable of explaining the creation of an environment of internal and external controls in the "field" of nature.[15] We understand now that man is not just a member of systems and metasystems (as a structural view would try to present it), but that he lives the processes of his life simultaneously at *all* levels because he shares in all of them in an integral way. The hierarchy of levels, in this perspective, is then no longer a causal (control) hierarchy, but a multilevel representation of self-transcendence; Gödel's theorem—that a formal system cannot represent itself (see Chapter 2 by Pankow)—is "circumvented" by this representation because there is always another level (another formal description) to relate to. As Fischer (1972) points out, Gödel's theorem refers only to the internal consistency of a particular state of consciousness, namely, the "normal state" of conscious learning with its single-level focus on the "I" (the worldly projection of the "Self"), of physical space-time and dualistic Aristotelian logic and language. It does not hold for the multilevel I–Self dialogue which is the domain of superconscious learning and in which cortical and subcortical activity ("mind" and "body") become indistinguishably enmeshed, representing the experience of oneness also at the physiological level. Spencer Brown's (1969, p. 105) warning that a self-reflective world "must first cut itself up into at least one state which sees, and at least one other state which is seen" and "in this condition . . . will always partially elude itself" refers to an inquiry into form and by formal means only. In this case, the world's "particularity is the price we pay for its visibility [p. 106]"; self-transcendence (the "desire to see") remains "the original mystery." But natural, in contrast to formal, division results in parts which are wholes in themselves (see Chapter 2 by Pankow), just as each cell contains the genetic program for the whole organism (and perhaps even more than that?) and each chip of a holographic plate yields the whole picture. In principle, a gestalt system can "see" the whole universe and the price paid in this case is poor resolution. It is through this gestalt perception that a self-transcendent system as a whole relates to its environment as a whole. *Mind*, in this perspective, may then "be regarded as the software of not

[15] Smuts's early recognition of "a universe of *whole*-making, not of soul-making" marks a historic breakthrough of process thinking which facilitates the overcoming of the structural system paradigms which Western thought has produced, e.g., panpsychism, monadism (the Spirit in everything), and Spinoza's philosophy, which views all bodies as animate.

ISBN 0-201-03438-7/0-201-03439-5pbk.

just the brain but of the *whole* self-organizing system, the living system" (Fischer, 1975a). It is in this sense that, through the interconnectedness of open systems, the universe may be said to be mind and that, as Huxley (1954) put it, "each one of us is potentially Mind-at-Large."

It would thus be wrong to say that evolutionary processes "build" a hierarchy from the bottom. Rather, we may say that system levels "mature" in a certain order, orchestrating evolution ever more fully. "Maturing" here means gaining flexibility in adaptation and action of the system "phenotype." Primitive man was no less a part of the mankind learning process than we are today and, in a way, he had culture. Every bioorganism, every atom which ever existed, shared in the universal evolution. But there is a "phylogeny" at every system level, a maturing process of the system "genotype" which seems to be characterized by the increasing importance of epigenetic development, for example, by cross-cultural transmission (Waddington, 1975, pp. 294f.). Whereas for animals, the social and especially the cultural system level are still immature, (i.e., fairly rigid), man has succeeded in maturing his social and partially also his cultural system "genotype" by means of superior learning modes. Evolution thus produces not just increasingly flexible and adaptive bioorganismic phenotypes (see Chapter 1 by Waddington), but also *increasingly flexible "phenotypes" of social systems and cultures.* At the same time, and even complementary to it, linear genetic development is gradually overtaken by *epigenetic development through intersystemic exchange*, emphasized already in higher animals and dominant in human sociogenesis, and even more so in noogenesis—until, in our time, mankind is setting out to reflect on itself for the first time in a global and total way. Hand in hand with this metaevolution goes a vast increase in indeterminacy or, as we may now call it, open potential, free will—or simply openness.

4. OTHER DESCRIPTIONS: COMPLEMENTARITY

A hierarchical description, open-ended in the upward and downward direction, embraces both aspects of evolution, emergence and purpose. These two aspects are *complementary.* For the human world, this may be expressed as follows: Viewed from one side, subjective planning tries to align normative action with emergence, and viewed from the other side, objective emergence in the human world is acted out in the framework of purposeful plans. I call this integral process, taking both points of view, *design for evolution* (Jantsch, 1975). It embraces the basic complementarity between determinism and free will, necessity and

ISBN 0-201-03438-7/0-201-03439-5pbk.

chance, objective and value-sensitive method (see also Chapter 9 by Marney and Schmidt).

All unfolding of energy may be described in dialectic terms (tension field between opposites) if viewed from outside, and in complementary terms (opposites containing each other) if viewed from inside. This may be understood as a nonequilibrium principle inherent in any historical process. Since we participate in evolution as its integral agent, a complementary description appears fitting for many purposes. Evolution in the human world may then be generally understood as the unfolding of energy in manifold manifestations, from physical (physical energy, matter, complexity) through social (emotions, motivation, social systems) to cultural/spiritual (knowledge, values, images of cosmos/man) manifestations and beyond.

A basic complementarity between psychic and physical manifestations of reality, the basis of many ancient myths, is again emphasized in our day by Jungian thought. Related to it is the suspected complementarity between timeless acausal orderedness and sporadic-temporal manifestations of psychophysical synchronicity (Franz, 1974, pp. 198f.). The same basic complementarity may be viewed as arising from an infinite regression of process levels within and without man, leading to the ultimate essence of reality in both directions, *atman* within and *brahman* without in terms of Vedanta—which are then recognized as identical. Thus, what ultimately emerges is a structural metaphor, also often expressed as a double spiral in mystic and artistic conceptions throughout the ages (see, e.g., Purce, 1974). A double spiral is also suitable for representing the development of human consciousness by means of the four learning modes, as outlined earlier in this chapter. But what counts in all these metaphors is not so much the structure, but the process of search within and without, often also represented by a spiral returning to itself, with no end points.

Learning, in the framework of the *atman–brahman* metaphor, may then proceed as "outer learning" through scientific method and as "inner learning" through experience and direct insight. The different functions of the two hemispheres of the human brain and the conscious and superconscious learning modes, as outlined earlier, are related to these concepts. Another expression of the complementarity of inner and outer learning may be recognized in the already mentioned complementarity between internal (coordinative) and external (competitive/Darwinian) factors of selection in evolution (Whyte, 1965). As Table 3.1 suggests, these are aspects relative to the level at which the evolutionary process is considered. In a multilevel view, the evolutionary process is both internal and external.

ISBN 0-201-03438-7/0-201-03439-5pbk.

A process view would also emphasize complementary notions which are obtained when looking at the process from different ends. Stochastic and teleological descriptions form such a pair, differentiation and integration another (see also Chapter 12 by Jantsch). Granit (1975) has found that in the development of the nervous system "purposiveness and adaptability are . . . partners in an indissoluble marriage between the organism and its environment" and that "adaptability is a teleological concept, with or without awareness of goal," a complementarity he terms *immanent teleology*. Every process, natural or machine-like, may be described in complementary causal and finalistic terms, where the finalistic description is often expressed through a variational principle, maximizing some factor or complex set of factors at every stage of the stochastic process. In evolution, however, the variational principle itself has probably to be assumed as evolving, perhaps in accordance with the evolving modes of learning.

In an evolutionary process, every movement is the seed of its own seed, or, as the Sufi mystic Ibn'Arabi put it, "each cause is the effect of its own effect." Life can only come from life. In modern physics, such a process view has become known as the "bootstrap model" (Chew, 1968), which holds that nature cannot be reduced to fundamental entities (building blocks of matter, laws, or pinciples), but evolves entirely through the self-consistency of a web of processes. In the subnuclear domain, all hadrons (strongly interacting particles) may become constituents of each other as well as part of the binding forces exchanged between the constituents. Thus, hadrons continuously generate each other and the whole set of hadrons "pulls itself up by its own bootstrap" (see also, in Chapter 2, Pankow's evocation of Baron Münchhausen as the symbol for self-transcendence). Such a microscopic view of interconnected evolving relations has been expressed in more general terms as a universal paradigm in Buddhism (Capra, 1976). This paradigm may now be called "self-realization through self-transcendence."

For some purposes, however, it is of advantage to dissolve the complementarity of emergence and purpose in order to focus on *either* end, on stochastic or teleological formulation. In the scheme of Table 3.1, this would amount to focusing on the upward or downward links in the learning processes.

A nonequilibrium approach focusing on *emergence* would attempt to describe how systems live out their immanent potential, the process values inherent in their proper level. Such intrinsic core values may perhaps be expressed in the following terms:

- *Physical systems:* Maintenance and renewal of capability for entropy production (energy exchange with the environment);

ISBN 0-201-03438-7/0-201-03439-5pbk.

- *Bioorganisms:* Flexibility of phenotype;
- *Ecosystems* (as functionally learning social systems): Resilience (see Chapter 4 by Holling);
- *Social systems:* Mutual correlation between social system and socioecological environment;
- *Cultures:* Coherence and viability of cultural (cosmological, mythological, philosophical, ethical, etc.) paradigms in a changing world;
- *Mankind:* Being "gardener of the planet Earth" (de Jouvenel, 1968) in a total sense, from physical to spiritual aspects.

What seems to come into focus here is always the correlation between a system and its proper environment, or in other words, the continuity of energy exchange at many levels. If evolution is the unfolding of energy at many levels, *self-transcendent systems are the means to ensure the continuity of evolution, and vice versa.*

Perhaps the most important principle underlying this mutual multilevel correlation between system and environment is *symbiotization* (see also Chapter 10 by Maruyama). Although a comprehensive theory of this basic phenomenon is still lacking (for a promising approach to its exploration, see Chapter 7 by Zeleny and Pierre), there is increasing evidence that much of the living world is organized also as a symbiotic ecosystem and not just as an organismic system. Perhaps it would be more correct to say that *horizontal ecosystemic (symbiotic) and vertical organismic organization are complementary aspects of all biological, and possibly all living, systems.* The total energy conversion and exchange of the biological world seems to be managed between two kinds of organelles, the mitochondria in animals and the chloroplasts in plants. They resemble genetically independent bacteria which, in the course of evolution, became engulfed by larger cells and now live in symbiosis with the rest of the organism. There is even a recent encouraging view (Thomas, 1974) that most approaches between living systems constitute basically friendly invitations to join in symbiosis and that many pathological system states such as illness may be explained in terms of misunderstanding and unwarranted paranoia—the system overreacting and releasing all its defenses at once, and thereby often hurting itself most. The defensive survival syndrome seems to impair the basic trust and openness of life to some extent at the sociobiological level, but even much more so at the sociocultural level. "The only principle which does not inhibit progress is: *anything goes*"; Feyerabend's (1975) prescription for science may also point to what is needed most for many more aspects of sociocultural life.

Generally, stochastic formulations at any level—physical, social, cultural—describe processes of *differentiation.* What may be studied, in particular, are the relations between energy exchange across system boundaries, fluctuations, and structural integrity. Chapters 4 by

ISBN 0-201-03438-7/0-201-03439-5pbk.

Holling, 5 by Prigogine, and 10 by Maruyama explore this perspective at different learning levels.

But with systems of interactive populations and processes—systems with history—we deal with space-time structures and therefore also with coincidences, or synchronicity both in a general sense and in the specific psychophysical meaning which Jung has given this term. Waddington's concept of "chreods" traversing the "epigenetic landscape" (see, e.g., Waddington, 1975, pp. 258ff.) is useful here, as are other descriptions in terms of a *field* or a vibrating "spacework" (see Chapter 6 by Abraham).

Looking from the other end of the process, a teleological description focuses on the *integration* of a system concept at a given level with the values inherent in the adjacent levels, especially in the superordinate level of a hierarchical structure. A nonequilibrium process-oriented approach would then try to vindicate any step taken in evolutionary experimentation by virtue of its contribution to a higher-order process. In this way, a *process teleology* builds up. The evolution of human beings may then find its goal in the evolution of social systems which, in turn, find their goals in the evolution of cultures, and so forth, over the evolution of mankind to Teilhard de Chardin's "noogenesis" and universal evolution, and possibly even beyond. There is no need ever to come to an end, which would signify a permanent structure of some kind. In different categories, such a process teleology may also be expressed in terms of a *learning hierarchy:* Virtual learning finds its goal in the development of functional learning, which in turn makes possible the development of conscious learning, and so forth to superconscious learning and beyond. As higher forms of learning open up new degrees of freedom, they also introduce new constraints; where genetically programmed constraints limited and guided functional learning, morality does the same for conscious learning, and so forth in a constraint hierarchy. The emergent image is that of man-in-universe, or more aptly, of man-becoming-in-universe-becoming, the image which Huxley (1945) has termed the "perennial philosophy."

But a process teleology goes even further in making it possible to view human existence at all levels simultaneously by virtue of homologous relations across the levels. Social relations, cultural values, and mankind paradigms all result from processes expressing themselves through the individual. In the evolutionary conception of ancient Chinese philosophy, man lives simultaneously the three great taos, or right paths—the tao of heaven, the tao of man, and the tao of the earth. In a process view, there is no center. In fact, a process view is capable of describing man in terms of a multiple-personality metaphor which is suggested so strongly by experiences in superconscious learning (e.g., meditation and metanoia, leading to the experience of a new reality; see

ISBN 0-201-03438-7/0-201-03439-5pbk.

also Chapter 11 by Markley). Man may then be viewed as the multiple intersection of a multiplicity of processes (as in Buddhism, which knows no central self).

Since there is no center, it is also permissible to invert the process teleology and to suggest, for example, with Dunn (1971), that "human beings can establish the process of human development as the goal of the process of social evolution." Ultimately, universal evolution may be justified by the life of any single creature. Such an inversion leads back to the complementary description of the evolutionary process and to Pankow's statement in Chapter 2 that "parts are themselves wholes."

5. SOME CONSEQUENCES FOR HUMAN DESIGN

Human design includes the important and unique faculties of organization and evaluation. Organization may be applied to the stabilization of system structure as well as to the fostering of processes of evolutionary experimentation; the subsequent evaluation becomes the stepping-stone for a new conservative or innovative venture of organizing activity.

Much evidence presented in this volume suggests that organization *qua* stabilization may turn out to be disastrous in certain circumstances. Quite generally, equilibration tends to make an ecosystem nonresilient and unhealthy, and may ruin it (see Chapter 4 by Holling). Moreover, dissipative structures tend to switch to new dynamic regimes—usually involving structural change—in times of crisis when their energy exchange processes tend to become stifled (see Chapter 5 by Prigogine). At such times, which may also be characterized by an imbalance between the internal and external learning processes of a system, the principle of "order through fluctuation" ought to be furthered, not impaired by efforts to preserve the old structure. In fact, stabilization may be expected to stave off the mutation for a relatively short period only and invite a much more disruptive switch to a new regime. Evolutionary processes cannot be contained indefinitely in structures which are stabilized by external management. Life will eventually try to break through any artificial system boundaries. Two basic rules for purposeful organization or planning seem to follow immediately: (1) Make structure as flexible as possible, so that it can be changed by emergent changes in the interaction of processes; and (2) identify and enhance evolutionary processes of change, recognize and respect their self-bounding nature.

Learning makes the "hard" system/process hierarchy malleable in an ascending order. At the present stage, strategic planning is turning social system design into a more self-reflective exercise. It has already

ISBN 0-201-03438-7/0-201-03439-5pbk.

made physical planning, in particular the planning of technology, much more flexible and sensitive to the values of social systems, even if we are still far from the ideal. But industrial planning geared to the *function* of technology in society and no longer solely to technology's intrinsic values (feasibility, extremalization of single performance parameters, etc.), and the basic idea of Planning-Programming-Budgeting (PPBS) in government at various jurisdictional levels have led to a more conscious exploration of process options as well as to the incorporation of organizational flexibility as a design principle (Jantsch, 1972, 1975). But social systems and processes are still designed against the background of a rarefied structural *telos* engrained in Western culture and expressed through rigid institutional patterns and policies. Conscious learning alone, reaching up to the level of culture, cannot introduce or manage cultural flexibility; the top level, touched by a particular kind of learning, so far always appears as an immovable or capriciously changing authority which we try to please by ascribing permanency to it—as long as it does not evolve by itself, switching to a new regime.

To "soften up" the cultural level, too, is not a task for conscious, but for superconscious learning. It needs to be changed by new images emanating from the mankind level. To facilitate the coming of a new image of man is therefore the most reasonable attitude we can entertain in a time of impending major cultural transformation (see Chapter 11 by Markley). In human history, such new images of man have often changed the cultural level and established regional cultures nourishing social development over millennia. But for the first time, we have now the opportunity to use conscious learning to further their emergence, in other words, to consciously accept and improve superconscious learning. An increasing number of people, especially among the young, seems ready for this new task which clashes with established norms of Western-style education.

But we do not yet see very clearly what the joining of conscious and superconscious learning may mean at this moment in mankind's evolution. Laszlo (1975) has made the interesting point that, so far, only human individuals are capable of acting as *proactive* systems, extrapolating the possible transformations as well as anticipating the dangers which make them necessary. In contrast, sociocultural systems are of an *anticipatory-reactive* type, anticipating changing conditions as related to the limits of the existing negative feedback stabilizing functions; only if the conditions head toward the limits of the error-correcting thresholds of the existing system, are internal forces (reform and revolution) activated which force the system toward transformation. But mankind as a global system still belongs to an almost nonan-

ISBN 0-201-03438-7/0-201-03439-5pbk.

ticipatory *reactive* type, renorming its transitory stable states only when the environmental inputs have reached or surpassed the error-correction thresholds of their existing states of stability. If this scheme continues to hold during the forthcoming macrotransformation, mankind would stand little chance of influencing the crisis. What would be needed is a much improved coupling of superconscious with fully developed conscious learning to elevate this hierarchy and at least put mankind on an anticipatory-reactive, and social systems on a proactive, level. But maybe we have not yet matured to the point where we are "supposed" to interfere effectively; maybe we learn not just by doing, but also by being done-by—as the macrotransformation happens, we begin to understand it. We must not forget that beyond the cultural environment the correlation between man and his "mankind environment" is just starting now. We may have brutal lessons yet to learn.

The same dilemma is put into even sharper focus by the following observation: As evolutionary self-reflection moves to higher levels, changes at the lower levels tend to occur at higher frequencies. In the human world, physical (technological) change runs at higher frequencies than social change, which, in turn, occurs more frequently than cultural change, which is still more frequent than the evolutionary thresholds crossed by the mankind process. Such acceleration of change at the lower levels reflects the emergence of a *control hierarchy;* quite generally, for a hierarchy to be controllable in cybernetic terms, the lower level must act at a higher frequency than the upper one (Mesarović et al., 1970, p. 50). Thus, as the human genotype correlates with higher levels of consciousness and environment, his increasing flexibility in acting and introducing change tends to express itself primarily at the lower levels of the control hierarchy which come within his grasp. But this means that all his intellectual and technological power, the thrust of his activities as well as his ambitions, are ultimately controlled—not by himself, as he likes to believe, but by cultural and mankind processes beyond his conscious understanding. Our only chance to understand these processes better is to listem to them better.

But I believe also that many of the problematic developments in contemporary society are due to the growth in size, scope, and number of such structure-oriented organismic control hierarchies. Not long ago, they still seemed to present the only rationale for dealing with large, richly joined systems facing a turbulent environment (Emery and Trist, 1965). But more recently, it is becoming clear that an emerging macro-organization theory will have to include such phenomena as differentiation, pluralism, symbiosis, structural change, and evolving policies also at the macrolevel (Metcalfe, 1974; see also Chapter 10 by Maruyama). Elsewhere (Jantsch, 1975), I have argued that we may soon see a dy-

ISBN 0-201-03438-7/0-201-03439-5pbk.

namic play of cultural pluralism unfolding on the substrate of a more uniform social fabric, thus reversing the traditional control hierarchy. "Life-style laboratories" such as Berkeley, California, or Boulder, Colorado, seem to herald such an enhanced balance among the levels of evolutionary self-attunement. With it, the semiautonomous thrust of technological change may gradually subside and a greater flexibility become introduced at all levels simultaneously. Of course, such a multilevel process can then no longer be controlled in the same strict cybernetic terms—it has simply to be *lived.*

Another contemporary aspect of human evolution seems to call for a new kind of balance: Viewed from a certain angle, the process of alignment with the evolution seems to become ever more precarious. A particular instance of this is the possibility of increasing *metastability* due to the multivariate and flexible coupling of subsystems in human societies (see Chapter 5 by Prigogine). Under conditions of metastability (a notion which seems here intricately related to Holling's concept of resilience—see Chapter 4), a system can absorb bigger fluctuations than under normal stable conditions, without switching to a new regime. For the "normal," stepwise evolution of a system this would be a double-edged sword: on the one hand, a major shake-up such as Arab oil price policy since late 1973 introduced major instabilities, but did not basically change the structure of Western economy and of most of its relations with the rest of the world; but on the other hand, this seems to now invite bigger and bigger fluctuations—a new Middle East war? A new oil embargo? New price cartels? A "forward escape" in the form of a Western takeover of non-Western oil resources? At present all the foregoing remain openly discussed possibilities until things change more profoundly.

A paradox seems to pose itself here: Living near boundaries of a stable domain (i.e., in sufficient nonequilibrium) is a precondition for a system's capability of evolving, of changing to another, qualitatively different stable domain (Prigogine); but at the same time, apparently through more effective flexible coupling of the subsystems (Prigogine), it enhances the system's resilience, or capability for persistence in the original domain (Holling), which, in turn, tends to increase the fluctuations until they are big enough to drive the system to a new regime (Prigogine). Does life with high fluctuations enhance *both* structural persistence and transformability? And at what point and by what mechanism is structural persistence relinquished in favor of process persistence, or long-term viability through qualitative transformation of the space-time structure? This crucial question has not yet found an answer for complex sociobiological and sociocultural systems. But it would not come as a surprise if the capabilities of resisting well and of transform-

ISBN 0-201-03438-7/0-201-03439-5pbk.

ing well turned out to be just complementary aspects of that same systems' flexibility which we keep encountering in so many manifestations. After all, is this not what is implied by a life lived to the fullest extent?

But can flexibility itself evolve further, enter a new regime, when the fluctuations invited by it assume threatening dimensions? What is the meaning of flexibility and resilience (or metastability) in a human world which has itself prepared technological "fluctuations" which might throw much of the planetary life into extinction and change the environment permanently in a sense inimical to life?

With other contributors to this volume, I believe that we find ourselves at an evolutionary threshold, marked by a new kind of *dynamic balance*, of "going with the fluctuations" and thereby "defusing" them, of near-continuous qualitative change and cultural pluralism. Rather than the conscious design of specific cultural regimes and smooth transitions toward them, the beginnings of *cultural design*, or the conscious adaptation of the cultural environment to man—the next step in the correlation between man and his multilevel environment—will be characterized by the design of a *metaregime* of balanced transformations, more or less in permanence. Such a metaregime may also signify the unconscious beginnings of man's correlation with a yet further level of environment, namely, mankind-at-large.

For the time being, a dynamic balance of "going with the fluctuations" may have started without our endorsement. Inflation in the Western world may actually be a healthy process, adjusting the gross imbalance between the material standards of living in the industrialized and in the developing countries—an adjustment which we accept without too much grumbling, because it is based not on a moral decision (on human design), but on "circumstances beyond control." The same may be said for a general adjustment of trade relations between raw material producing and industrialized countries, which still has to be forced by unilateral coercion. We are still missing out on most of our chances for creative design—maybe being controlled by cultural and mankind processes beyond our grasp is what we subconsciously prefer, after all.

There is obviously an optimal design principle to be explored in the joining of systems. Too loose coupling would mean that fluctuations (innovations) emanating from one system would not reach others; too close coupling would mean that the environment can quickly damp out any fluctuation. To become creative, human society has to be organized horizontally in flexibly joined, pluralistic, and possibly symbiotic systems at "optimal distance" from each other, and vertically in a balanced interplay of learning modes at the corresponding levels of human

ISBN 0-201-03438-7/0-201-03439-5pbk.

existence. The emerging world system must not become a control hierarchy with a strong world government on top to maintain distorted relations, but a healthy web of processes lived out at many levels simultaneously, balancing individual, social, cultural, and mankind aspects of all expressions of human life.

REFERENCES

Assagioli, Roberto (1971). *Psychosynthesis.* New York: Viking Compass Book.

Bateson, Gregory (1972). *Steps to an Ecology of Mind.* San Francisco: Chandler; New York: Ballantine paperback.

Capra, Fritjof (1976). *The Tao of Physics: An Exploration of the Parallels between Modern Physics and Eastern Mysticism.* Berkeley, Calif.: Shambhala.

Chew, Geoffrey F. (1968). "Bootstrap: A Scientific Idea?," *Science,* 161, 762.

Dunn, Edgar S., Jr. (1971). *Economic and Social Development: A Process of Social Learning.* Baltimore and London: Johns Hopkins Press.

Eigen, Manfred (1971). "Self-Organization of Matter and the Evolution of Biological Macromolecules," *Naturwissenschaften,* 58, 465-522.

Eigen, Manfred, and Winkler, Ruthild (1975). *Das Spiel: Naturgesetze steuern den Zufall.* Munich and Zurich: Piper.

Emery, Fred E., and Trist, E. L. (1965). "The Causal Texture of Organizational Environments," *Human Relations,* 18, 21-32.

Feyerabend, Paul (1975). *Against Method.* London: NLB; Atlantic Highlands, N.J.: Humanities Press.

Fischer, Roland (1966). "Biological Time," in *The Voices of Time* (J. T. Fraser, ed.). New York: Braziller.

Fischer, Roland (1971). "A Cartography of the Ecstatic and Meditative States," *Science,* 174, 897-904.

Fischer, Roland (1972). "On Separateness and Oneness," *Confin. psychiat.,* 15, 165-194.

Fischer, Roland (1975a). "Mind: The Software of the Brain," *ASC Cybernetics Forum,* VIII, 15-16.

Fischer, Roland (1975b). "Cartography of Inner Space," in *Hallucinations: Behavior, Experience and Theory* (Ronald K. Siegel and Louis J. West, eds.). New York: Wiley.

Fischer, Roland, and Rhead, John (1974). "The Logical and the Intuitive," *Main Currents,* 31, 50-54.

Fontela, Emilio, and Gabus, André (1974). "Dematel: Progress Achieved," *Futures,* 6, No. 4, 361-363.

Franz, Marie-Louise von (1974). *Number and Time: Reflections Leading toward a Unification of Depth Psychology and Physics* (A. Dykes, transl.). Evanston, Ill.: Northwestern Univ. Press.

ISBN 0-201-03438-7/0-201-03439-5pbk.

Frisch, Karl von, and Frisch, Otto von (1974). *Animal Architecture*. New York: Harcourt.

Granit, Ragnar (1975). "Adaptability of the Nervous System and Its Relation to Chance, Purposiveness and Causality," in: *Science and Absolute Values*, Proceedings of the Third International Conference on the Unity of the Sciences, Tarrytown, N.Y.: International Cultural Foundation.

Grof, Stanislav (1975). *Realms of the Human Unconscious: Observations from LSD Research*. New York: Viking.

Horn, G., Rose, S. P. R., and Bateson, P. P. G. (1973). "Experience and Plasticity in the Central Nervous System," *Science*, 181, 506-514.

Huxley, Aldous (1945). *The Perennial Philosophy*. New York: Harper.

Huxley, Aldous (1954). *The Doors of Perception*. New York: Harper and Row.

Illich, Ivan (1975). "Spezifische Kontraproduktivität," lecture given at the Federal Polytechnical Institute (ETH), Zurich, Switzerland.

Jantsch, Erich (1972). *Technological Planning and Social Futures*. London: Associated Business Programmes; New York: Halsted Press.

Jantsch, Erich (1975). *Design for Evolution: Self-Organization and Planning in the Life of Human Systems*. New York: Braziller.

Jouvenel, Bertrand de (1968). *Arcadie, essais sur le mieux vivre*. Paris: SEDEIS.

Jung, Carl Gustav (1949). Foreword to the *I Ching*, The R. Wilhelm translation rendered into English by C. F. Baynes. Princeton, N.J.: Princeton Univ. Press, 1950 (3d ed. 1967).

Laszlo, Ervin (1972). *Introduction to Systems Philosophy: Toward a New Paradigm of Contemporary Thought*. New York: Gordon and Breach; Harper Torchbooks.

Laszlo, Ervin (1975). "Goals for Global Society—A Positive Approach to the Predicament of Mankind," in: *Science and Absolute Values*, Proceedings of the Third International Conference on the Unity of the Sciences; Tarrytown, N.Y.: International Cultural Foundation.

Meadows, Donella H., Meadows, Dennis L., Randers, Jörgen, and Behrens, William W., III (1972). *The Limits to Growth*. New York: Universe Books.

Mesarović, Mihajlo D., Macko, D., and Takahara, Y. (1970). *Theory of Hierarchical, Multilevel Systems*. New York: Academic Press.

Mesarović, Mihajlo D., and Pestel, Eduard (1974). *Mankind at the Turning Point: The Second Report to the Club of Rome*. New York: Dutton.

Metcalfe, J. Les (1974). "Systems Models, Economic Models and the Causal Texture of Organizational Environments: An Approach to Macro-Organization Theory," *Human Relations*, 27, 639-663.

Naranjo, Claudio (1972). *The One Quest*. New York: Viking.

Ozbekhan, Hasan (1976). "The Predicament of Mankind," in: *World Modeling: A Dialogue* (C. West Churchman and Richard O. Mason, eds.). Amsterdam and Oxford: North-Holland; New York: American Elsevier.

Prigogine, Ilya (1973). "Irreversibility as a Symmetry Breaking Factor," *Nature*, 248, 67-71.

ISBN 0-201-03438-7/0-201-03439-5pbk.

Purce, Jill (1974). *The Mystic Spiral.* London: Thames and Hudson; New York: Avon Books.

Sackman, Harold (1967). *Computers, System Science, and Evolving Society.* New York: Wiley.

Smuts, Jan Christiaan (1926). *Holism and Evolution.* Republished New York: Viking (1961).

Spencer Brown, G. (1969). *Laws of Form.* London: Allen and Unwin; New York: Julian Press (1972); Bantam paperback (1974). Page numbers refer to the Bantam edition.

Thomas, Lewis (1974). *The Lives of a Cell.* New York: Viking; Bantam paperback (1975).

Vickers, Geoffrey (1973). *Making Institutions Work.* London: Associated Business Programmes; New York: Halsted Press.

Waddington, Conrad H. (1974). "The Mystery of the Libidinous Molecule," *The New York Review,* 28 November 1974.

Waddington, Conrad H. (1975). *The Evolution of an Evolutionist.* Edinburgh: Edinburgh Univ. Press; Ithaca, N.Y.: Cornell Univ. Press.

Whyte, Lancelot L. (1965). *Internal Factors in Evolution.* New York: Braziller.

Wilson, Edward O. (1975). *Sociobiology: The New Synthesis.* Cambridge, Mass.: Belknap Press of Harvard Univ.

ISBN 0-201-03438-7/0-201-03439-5pbk.

Part II

Formal Approaches to Evolving Systems

A plague was approaching Ephesus, and before the disease broke out, Apollonius was aware of its imminence, and foretold it correctly. Often in his discourse he would say, "Earth, stay as you are" and, in threatening tones, such things as "Save these people" and "You are not to come here."

Philostratus, *Life of Apollonius,* IV

Self-transcendent systems can never be fully represented through formal approaches. No formal model will ever be adequate to describe all facets of life, and in particular the complex expressions of human life. However, taking a process-oriented view and treating self-transcendent systems in a formal way as open systems far from equilibrium and capable of changing their dynamic regimes admits already of analogous representation in terms of that basic characteristic by which they distinguish themselves most sharply from equilibrium and machine-like structures, namely, coherent behavior or "continuity in change." A decisive advantage is thereby gained over the habitual structure- and equilibrium-oriented mechanistic and behavioral models, which are not capable of representing qualitative change.

The empirically derived concept of *nonequilibrium ecosystems* (Holling) focuses on the notion of resilience, or a system's capability to persist in a globally stable dynamic regime if it is far from equilibrium. The concept of "*order through fluctuation*" (Prigogine) goes further in describing how nonequilibrium systems, through internal amplification of fluctuations, may cross instability thresholds and form so-called dis-

ISBN 0-201-03438-7/0-201-03439-5pbk.

Erich Jantsch and Conrad H. Waddington (eds.), *Evolution and Consciousness: Human Systems in Transition.*

sipative structures, or new dynamic regimes which represent a new ordering principle beyond the thermodynamic branch (which describes order near equilibrium). Such dissipative structures can be maintained only through energy exchange with the environment—Being, for them, is continuous Becoming, the same condition found for self-transcendent systems and life in general. Process (or function) and structure, deterministic and stochastic features, necessity and chance (or free will), become complementary aspects in the self-organizing dynamics of "order through fluctuation" which may also be graphically depicted as a nonequilibrium system "stumbling forward" and crossing by its own force the ridges separating "valleys" of global stability.

A complementary approach, the *theory of catastrophe* (Thom/ Abraham), focuses on the existence of multiple globally stable regimes (called *macrons*, and equivalent to dissipative structures) and the transitions (catastrophes) between them. Macrons are, at the present stage of the theory, represented by mathematical descriptions of their equilibrium state (attractors). Therefore, catastrophes appear as sudden quantum jumps, as if due to "pushes" by an outside force, comparable to a golf ball being propelled over a ridge by a single stroke. What is of central interest in this approach, is the landscape of new forms, the "epigenetic landscape," beyond the ridge.

The kind of dynamic balance which implies qualitative pluralism and near-continuous change at the same time seems to require some additional description of morphogenesis through interactive processes and symbiosis as it richly unfolds in broad valleys which are separated by low ridges. Simulation by means of *autopoietic models*, or models of self-renewing systems (Zeleny and Pierre), seems to hold great potential for exploring aspects of this life by focusing on interactive processes rather than on structure and components. In this approach, what is modeled are not the processes themselves, but the underlying sets of instructions which are paraphrased by random simulation runs.

Thus, the central part of this volume constitutes an attempt at communicating encouraging, even exciting, news, as well as at formulating a staggering challenge: There exists a primitive body of first formal approaches toward dealing with the wholeness and self-expression of living systems; it seems possible, to some degree, to extend it so as to include also partial descriptions of how such systems transcend their own level of existence in order to join in a wholeness beyond their own. In other words, instead of imposing our science upon life, we can find ways to let life guide the development of knowledge about ourselves.

E. J.

ISBN 0-201-03438-7/0-201-03439-5pbk.

4

Resilience and Stability of Ecosystems

C. S. Holling

An equilibrium-centered view is essentially static and provides little insight into the transient behavior of systems that are not near the equilibrium. Natural, undisturbed systems are likely to be continually in a transient state; they will be equally so under the influence of man. As man's numbers and economic demands increase, his use of resources shifts equilibrium states and moves populations away from equilibria. The present concerns for pollution and endangered species are specific signals that the well-being of the world is not adequately described by concentrating on equilibria and conditions near them. Moreover, strategies based upon these two different views of the world might well be antagonistic. It is at least conceivable that the effective and responsible effort to provide a maximum sustained yield from a fish population or a nonfluctuating supply of water from a watershed (both equilibrium-centered views) might paradoxically increase the chance for extinctions.

The purpose of this chapter is to explore both ecological theory and the behavior of natural systems to see if different perspectives of their behavior can yield different insights useful for both theory and practice. In a final section, such new insights will be discussed with a view to what may be learned from them for the management of human systems.

1. SOME THEORY

Phase Space Representation of Systems Behavior

Let us first consider the behavior of two interacting populations: a predator and its prey, a herbivore and its resource, two competitors, or any flexibly coupled subsystems of an ecosystem. It becomes very difficult on time plots to show the full variety of responses possible, and it proves convenient to plot a trajectory in a phase plane which represents the sequential change of the two populations at constant time intervals.

ISBN 0-201-03438-7/0-201-03439-5pbk.

Erich Jantsch and Conrad H. Waddington (eds.), *Evolution and Consciousness: Human Systems in Transition.*

Each point represents the unique density of each population at a particular point in time and the arrows indicate the direction of change over time (Figure 4.1). If oscillations are damped, then the trajectory is represented as a closed spiral that eventually reaches a stable equilibrium.

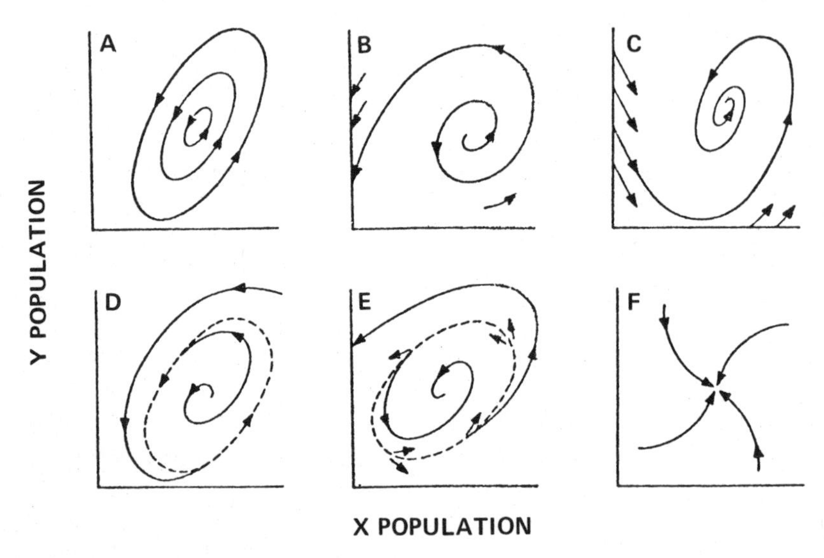

FIGURE 4.1. Examples of possible behaviors of systems in a phase plane; (a) neutrally stable cycles, (b) unstable equilibrium, (c) stable equilibrium, (d) stable limit cycle, (e) domain of attraction, (f) stable node.

We can imagine a number of different forms of trajectories in the phase plane. Figure 4.1b shows an open spiral which would represent situations where fluctuations gradually increase in amplitude. The small arrows are added to suggest that this condition holds no matter what combination of populations initiates the trajectory. In Figure 4.1a the trajectories are closed and, given any starting point, eventually return to that point. It is particularly significant that each starting point generates a unique cycle and there is no tendency for points to converge to a single cycle or point. This can be termed "neutral stability" and it is the kind of stability achieved by an imaginary frictionless pendulum. Figure 4.1c represents a stable system in which all possible trajectories in the phase plane spiral into an equilibrium. These three examples are relatively simple and, however relevant for classical stability analysis, may well be theoretical curiosities in ecosystems.

Figures 4.1d–f add some complexities. In a sense, Figure 4.1e represents a combination of parts b and c, with a region in the center of the phase plane within which all possible trajectories spiral inward to

ISBN 0-201-03438-7/0-201-03439-5pbk.

equilibrium. Those outside this region spiral outward and lead eventually to extinction of one or the other population. This is an example of local stability in contrast to the global stability of Figure 4.1c. I designate the region within which stability occurs as the *domain of attraction*, and the line that contains this domain as the *boundary* of the attraction domain.

The trajectories in Figure 4.1d behave just in the opposite way. There is an internal region within which the trajectories spiral out to a stable *limit cycle* and beyond which they spiral inward to it. Finally, a *stable node* is shown in Figure 4.1f, in which there are no oscillations and the trajectories approach the node monotonically. These six diagrams could be combined in an almost infinite variety of ways to produce several domains of attraction within which there could be a stable equilibrium, a stable limit cycle, a stable node, or even neutrally stable orbits. Although I have presumed a constant world throughout, in the presence of random fluctuations of parameters or of driving variables (Walters, 1971), any one trajectory could wander with only its general form approaching the shape of the trajectory shown (see also Chapter 5 by Prigogine). These added complications are explored later when we consider real systems. For the moment, however, let us review theoretical treatments in the light of the possibilities suggested in Figure 4.1.

Ecological Stability Theory: State of the Art

The present status of ecological stability theory is very well summarized in a number of analyses of classical models, particularly May's (1973) insightful analyses of the Lotka–Volterra model and its expansions, the graphical stability analyses of Rosenzweig (1971; Rosenzweig and MacArthur, 1963), and the methodological review of Lewontin (1969).

May (1973) reviews the large class of coupled differential equations expressing the rate of change of two populations as continuous functions of both. The behavior of these models results from the interplay between (a) stabilizing negative feedback or density-dependent responses to resources and predation, and (b) the destabilizing effects produced by the way individual predators attack and predator numbers respond to prey density (termed the functional and numerical responses, as in Holling, 1961). Various forms have been given to these terms; the familiar Lotka–Volterra model includes the simplest and least realistic, in which death of prey is caused only by predators, predation is a linear function of the product of prey and predator populations, and growth of the predator population is linearly proportional to the same product. This model generates neutral stability as in Figure 4.1a, but the assumptions are very unrealistic since very few com-

ISBN 0-201-03438-7/0-201-03439-5pbk.

ponents are included, there are no explicit lags or spatial elements, and thresholds, limits, and nonlinearities are missing.

These features have all been shown to be essential properties of the predation process (Holling, 1965, 1966) and the effect of adding some of them has been analyzed by May (1973). He points out that traditional ways of analyzing the stability properties of models using analytical or graphical means (Rosenzweig and MacArthur, 1963; Rosenzweig, 1971, 1972) concentrate about the immediate neighborhood of the equilibrium. By doing this, linear techniques of analysis can be applied that are analytically tractable. Such analyses show that with certain defined sets of parameters stable equilibrium points or nodes exist (such as Figure 4.1c), whereas for other sets they do not, and in such cases the system is, by default, presumed to be unstable, as in Figure 4.1b. May (1973), however, invokes a little-used theorem of Kolmogorov (Minorsky, 1962) to show that all these models have either a stable equilibrium point or a stable limit cycle (as in Figure 4.1d). Hence he concludes that the conditions presumed by linear analysis are unstable, and in fact must lead to stable limit cycles. In every instance, however, the models are globally rather than locally stable, limiting their behavior to that shown in either Figure 4.1c or 4.1d.

There is another tradition of models that recognizes the basically discontinuous features of ecological systems and incorporates explicit lags. Nicholson and Bailey (1935) initiated this tradition when they developed a model using the output of attacks and survivals within one generation as the input for the next. The introduction of this explicit lag generates oscillations that increase in amplitude until one or the other of the species becomes extinct (Figure 4.1b). Their assumptions are as unrealistically simple as Lotka's and Volterra's; the instability results because the number of attacking predators at any moment is so much a consequence of events in the previous generation that there are "too many" when prey are declining and "too few" when prey are increasing. If a lag is introduced into the Lotka–Volterra formulation (Wangersky and Cunningham, 1957), the same instability results.

The sense one gains, then, of the behavior of the traditional models is that they are either globally unstable or globally stable; that neutral stability is very unlikely; and that when the models are stable, a limit cycle is a likely consequence.

Many, but not all, of the simplifying assumptions have been relaxed in simulation models, and there is one example (Holling and Ewing, 1971) that joins the two traditions initiated by Lotka and Volterra and by Nicholson and Bailey and, further, includes more realism in the operation of the stabilizing and destabilizing forces. These modifications are described in more detail later; the important features account-

ISBN 0-201-03438-7/0-201-03439-5pbk.

ing for the difference in behavior result from the introduction of explicit lags, a functional response of predators that rises monotonically to a plateau, a nonrandom (or contagious) attack by predators, and a minimum prey density below which reproduction does not occur. With these changes, a very different pattern emerges that conforms most closely to Figure 4.1e. That is, there exists a domain of attraction within which there is a stable equilibrium; beyond that domain the prey population becomes extinct. Unlike the Nicholson and Bailey model, stability becomes possible, although in a limited region, because of contagious attack. (Contagious attack implies that, for one reason or another, some prey have a greater probability of being attacked than others, a condition that is common in nature—see Griffiths and Holling, 1969). The influence of contagious attack becomes significant whenever predators become abundant in relation to the prey, for then the susceptible prey receive the burden of attention, allowing more prey to escape than would be expected by random contact. This "inefficiency" of the predator allows the system to counteract the destabilizing effects of the lag.

If this were the only difference, the system would be globally stable, much as in Figure 4.1c. The instability of the prey to reproduce at low densities, however, allows some of the trajectories to cut this reproduction threshold, and the prey become extinct. This introduces a lower prey density boundary to the attraction domain and, at the same time, a higher prey density boundary above which the amplitudes of the oscillations inevitably carry the population below the reproduction threshold. The other modifications in the model, some of which have been touched on earlier, alter this picture in degree only. The essential point is that a more realistic representation of the behavior of interacting populations indicates the existence of at least one domain of attraction. It is quite possible, within this domain, to imagine stable equilibrium points, stable nodes, or stable limit cycles. Whatever the detailed configuration, the existence of discrete domains of attraction immediately suggests important consequences for the persistence of the system and the probability of its extinction.

Such models, however complex, are still so simple that they should not be viewed in a definitive and quantitative way. They are more powerfully used as a starting point to organize and guide understanding. It becomes valuable, therefore, to ask what the models leave out and whether such omissions make isolated domains of attraction more or less likely. Theoretical models generally have not done well in simultaneously incorporating realistic behavior of the processes involved, randomness, spatial heterogeneity, and an adequate number of dimensions or state variables. This situation is changing very rapidly as theory

ISBN 0-201-03438-7/0-201-03439-5pbk.

and empirical studies develop a closer technical partnership. In what follows I refer to real world examples to determine how elements that tend to be left out might further affect the behavior of ecological systems.

2. SOME REAL WORLD EXAMPLES FROM ECOLOGY

Self-Contained Ecosystems

In the broadest sense, the closest approximation we could make of a real world example that did not grossly depart from the assumptions of the theoretical models would be a self-contained system that was fairly homogeneous and in which climatic fluctuations were reasonably small. If such systems could be discovered, they would reveal how the more realistic interaction of real world processes could modify the patterns of systems behavior described earlier. Very close approximations to any of these conditions are not likely to be found, but if any exist, they are apt to be freshwater aquatic ones. Freshwater lakes are reasonably contained systems, at least within their watersheds; the fish show considerable mobility throughout, and the properties of the water buffer the more extreme effects of climate. Moreover, there have been enough documented man-made disturbances to liken them to perturbed systems in which either the parameter values or the levels of the constituent populations are changed. In a crude way, then, the lake studies can be likened to a partial exploration of a phase space of the sorts shown in Figure 4.1. Two major classes of disturbances have occurred: first, the impact of nutrient enrichment from man's domestic and industrial wastes, leading to eutrophication; and second, changes in fish population by harvesting. The overall pattern emerging from such studies is the sudden appearance or disappearance of populations, a wide amplitude of fluctuations, and the establishment of new domains of attraction.

The history of the Great Lakes provides not only some particularly good information on responses to man-made enrichment, but also on responses of fish populations to fishing pressure. Eutrophication experience can be viewed as an example of systems changes in driving variables and parameters, whereas the fishing example is more an experiment in changing state variables. The fisheries of the Great Lakes have always selectively concentrated on abundant species that are in high demand. Prior to 1930, before eutrophication complicated the story, the lake sturgeon in all the Great Lakes, the lake herring in Lake

ISBN 0-201-03438-7/0-201-03439-5pbk.

Erie, and the lake whitefish in Lake Huron were intensively fished (Smith, 1968). In each case, the pattern was similar: a period of intense exploitation during which there was a prolonged high-level harvest, followed by a sudden and precipitous drop in populations. Most significantly, even though fishing pressure was then relaxed, none of these populations showed any sign of returning to their previous levels of abundance. This is not unexpected for sturgeon because of their slow growth and late maturity, but it is unexpected for herring and whitefish. The maintenance of these low populations in recent times might be attributed to the increasingly unfavorable chemical or biological environment, but in the case of the herring, at least, the declines took place in the early 1920s before the major deterioration in environment occurred. It is as if the population had been shifted by fishing pressure from a domain with a high equilibrium to one with a lower one. This is clearly not a condition of neutral stability as suggested in Figure 4.1a since, once the populations were lowered to a certain point, the decline continued even though fishing pressure was relaxed. It can be better interpreted as a variant of Figure 4.1e, where populations have been moved from one domain of attraction to another.

Since 1940, there has been a series of similar catastrophic changes in the Great Lakes that has led to major changes in the fish stocks. The explanations for these changes have been explored in part, and involve various combinations of intense fishing pressure, changes in the physical and chemical environment, and the appearance of a foreign predator (the sea lamprey) and foreign competitors (the alewife and carp). Whatever the specific causes, it is clear that the precondition for the collapse was set by the harvesting of fish, even though during a long period there were no obvious signs of problems. The fishing activity, however, progressively reduced the resilience of the system, so that when the inevitable unexpected event occurred, the populations collapsed. If it had not been the lamprey, it would have been something else: a change in climate as part of the normal pattern of fluctuation, a change in the chemical or physical environment, or a change in competitors or predators. These examples again suggest distinct domains of attraction in which the populations forced close to the boundary of the domain can then flip over it. The important point is not so much how stable populations are within a domain, but how likely it is for the system to move from one domain into another and so persist in a changed configuration.

Certainly some systems that have been greatly disturbed have fully recovered their original state once the disturbance was removed. But in most instances, the recovery took place in open systems in which reinvasion is the key ingredient. These cases are discussed later in connec-

ISBN 0-201-03438-7/0-201-03439-5pbk.

tion with the effects of spatial heterogeneity. For the moment, I conclude that distinct domains of attraction are not uncommon even within self-contained systems. If such is the case, then further confirmation should be found from empirical evidence of the way processes which link organisms operate, for it is these processes that are the cause of the behavior observed. It has indeed been found that fecundity and mortality will generate domains of attraction, with each domain separated from others by characteristic extinction and escape thresholds (Holling, 1965). So long as the populations remain within one domain, they have a consistent and regular form of behavior. If they pass a boundary to the domain by chance or through intervention by man, then the behavior suddenly changes in much the way suggested from the field examples discussed earlier.

The Random World: Heterogeneity over Time

To this point, I have argued as if the world were completely deterministic. In fact, the behavior of ecological systems is profoundly affected by random events. As one example, for twenty-eight years there has been a major and intensive study of the spruce budworm and its interaction with the spruce–fir forests of Eastern Canada (Morris, 1963). There have been six outbreaks of the spruce budworm since the early 1700s (Baskerville, 1971), and between the outbreaks the budworm has been an exceedingly rare species. When the outbreaks occur there is major destruction of balsam fir in all the mature forests, leaving only the less susceptible spruce, the nonsusceptible white birch, and a dense regeneration of fir and spruce. The more immature stands suffer less damage and more fir survives. Between outbreaks, the young balsam grow, together with spruce and birch, to form dense stands in which the spruce and birch, in particular, suffer from crowding. This process evolves to produce stands of mature and overmature trees with fir a predominant feature.

This is a necessary, but not sufficient, condition for the appearance of an outbreak; outbreaks occur only when there is also a sequence of unusually dry years (Wellington, 1952). Until this sequence occurs, it is argued (Morris, 1963) that various natural enemies with limited numerical responses maintain the budworm populations around a low equilibrium. If a sequence of dry years occurs when there are mature stand of fir, the budworm populations rapidly increase and escape the control by predators and parasites. Their continued increase eventually causes enough tree mortality to force a collapse of the populations and the reinstatement of control around the lower equilibrium. In brief, between outbreaks the fir tends to be favored in competition with spruce

ISBN 0-201-03438-7/0-201-03439-5pbk.

and birch, whereas during an outbreak spruce and birch are favored because they are less susceptible to budworm attack. This interplay with the budworm thus maintains the spruce and birch, which otherwise would be excluded through competition. The fir persists because of its regenerative powers and the interplay of forest growth rates and climatic conditions that determine the timing of budworm outbreaks.

This behavior could be viewed as a stable limit cycle with large amplitude, but it can be more accurately represented by a distinct domain of attraction determined by the interaction between budworm and its associated natural enemies, which is periodically exceeded through the chance consequence of climatic conditions. If we view the budworm only in relation to its associated predators and parasites, we might argue that it is highly unstable in the sense that populations fluctuate widely. But these very fluctuations are essential features that maintain persistence of the budworm, together with its natural enemies and its host and associated trees. By so fluctuating, successive generations of forests are replaced, assuring a continued food supply for future generations of budworm and the persistence of the system.

Until now, I have avoided formal identification of different kinds of behavior of ecological systems. The more realistic situations like budworm, however, make it necessary to begin to give more formal definition to their behavior. It is useful to distinguish two kinds of behavior. One can be termed *stability*, which represents the ability of the system to return to an equilibrium state after a temporary disturbance; the more rapidly it returns and the less it fluctuates, the more stable it would be. But there is another property, termed *resilience*, that is a measure of the persistence of systems and of their ability to absorb change and disturbance and still maintain the same relationships between populations or state variables. In this sense, the budworm forest community is highly unstable and it is because of this instability that it has an enormous resilience. I return to this view frequently throughout the remainder of this chapter.

In summary, many examples of the influence of random events upon natural systems further confirm the existence of domains of attraction. Most importantly, they suggest that instability, in the sense of large fluctuations, may introduce a resilience and a capacity to persist. They point out the very different view of the world that can be obtained if we concentrate on the *boundaries* to the domain of attraction rather than on equilibrium states. Although the equilibrium-centered view is analytically more tractable, it does not always provide a realistic understanding of the systems' behavior. Moreover, if this perspective is used as the exclusive guide to the management activities of man, exactly the reverse behavior and result can be produced than is expected.

ISBN 0-201-03438-7/0-201-03439-5pbk.

Spatial Heterogeneity

The final step is now to recognize that the natural world is not very homogeneous over space, as well, but consists of a mosaic of spatial elements with distinct biological, physical, and chemical characteristics that are linked by mechanisms of biological and physical transport.

The study of a peninsula on Vancouver Island, in which the topography and climate combine to make a mosaic of favorable locales for the tent caterpillar, provides a realistic example of the effects of both temporal and spatial heterogeneity of a population in nature (Wellington, 1964, 1965). From year to year, the size of these locales enlarges or contracts, depending on climate; Wellington was able to use the easily observed changes in cloud patterns in any year to define these areas. The tent caterpillar, to add a further element of realism, has identifiable behavioral types that are determined not by genetics but by the nutritional history of the parents. These types represent a range from sluggish to very active, and the proportion of types affects the shape of the easily visible web the tent caterpillars spin. From these defined differences of behavior with observations on changing numbers, shape of webs, and changing cloud patterns, an elegant story of systems behavior emerges: In a favorable year, locales that previously could not support tent caterpillars now can, and populations are established through invasion by the vigorous dispersers from other locales. In these new areas, they tend to produce another generation with a high proportion of vigorous behavioral types. Because of their high dispersal behavior and the small area of the locale in relation to its periphery, they tend to leave in greater numbers than they arrive. The result is a gradual increase in the proportion of more sluggish types to the point where the local population collapses. But although its fluctuations are considerable, even under the most unfavorable conditions there are always enclaves suitable for the insect. It is an example of a population with high fluctuations that can take advantage of transient periods of favorable conditions and that has, because of this variability, a high degree of resilience and capacity to persist.

3. SYNTHESIS

Some Definitions

Traditionally, discussion and analyses of stability have essentially equated stability with systems behavior. In ecology, at least, this has caused confusion since, in mathematical analyses, stability has tended

ISBN 0-201-03438-7/0-201-03439-5pbk.

to assume definitions that relate to conditions very near equilibrium points. This is a simple convenience dictated by the enormous analytical difficulties of treating the behavior of nonlinear systems at some distance from equilibrium. On the other hand, more general treatments have touched on questions of persistence and the probability of extinction, defining these measures as aspects of stability as well. To avoid this confusion, I propose that the behavior of ecosystems could well be defined by two distinct properties: resilience and stability.

Resilience determines the persistence of relationships within a system and is a measure of the ability of this system to absorb changes of state variables, driving variables, and parameters, and still persist. In this definition, resilience is the property of the system and persistence or probability of extinction is the result. *Stability*, on the other hand, is the ability of a system to return to an equilibrium state after a temporary disturbance. The more rapidly it returns, and with the least fluctuation, the more stable it is. In this definition, stability is the property of the system and the degree of fluctuation around specific states the result.

Resilience versus Stability

With these definitions in mind, a system can be very resilient and still fluctuate greatly, that is, have low stability. I have touched in the foregoing on the examples of the spruce budworm forest community in which the very fact of low stability seems to introduce high resilience. The balance between resilience and stability here is clearly a product of the evolutionary history of these systems in the face of the range of random fluctuations they have experienced.

In Slobodkin's (1964) terms, evolution is like a game, but a distinctive one in which the only payoff is to stay in the game. Therefore, a major strategy selected is not one maximizing either efficiency or a particular reward, but one which allows persistence by maintaining flexibility above all else. A population responds to any environmental change by the initiation of a series of physiological, behavioral, ecological, and genetic changes that restore its ability to respond to subsequent unpredictable environmental changes. Variability over space and time results in variability in numbers, and with this variability the population can simultaneously retain genetic and behavioral types that can maintain their existence in low populations together with others that can capitalize on chance opportunities for dramatic increase. The more homogeneous the environment in space and time, the more likely is the system to have low fluctuations and low resilience. It is not surprising, therefore, that the commercial fishery systems of the Great

ISBN 0-201-03438-7/0-201-03439-5pbk.

Lakes have provided a vivid example of the sensitivity of ecological systems to disruption by man, for they represent climatically buffered, fairly homogeneous and self-contained systems with relatively low variability and hence high stability and low resilience. Moreover, the goal of producing a maximum sustained yield may result in a more stable system of reduced resilience.

Nor is it surprising that, however readily fish stocks in lakes can be driven to extinction, it has been extremely difficult to do the same to insect pests of man's crops. Pest systems are highly variable in space and time; as open systems they are much affected by dispersal and therefore have a high resilience. Similarly, some Arctic ecosystems thought of as fragile may be highly resilient, although unstable. Certainly, this is not true for some subsystems in the Arctic, such as Arctic frozen soil, self-contained Arctic lakes, and cohesive social populations like caribou, but these might be exceptions to a general rule.

The notion of an interplay between resilience and stability might also resolve the conflicting views of the role of diversity and stability of ecological communities. Elton (1958) and MacArthur (1955) have argued cogently from empirical and theoretical points of view that stability is roughly proportional to the number of links between species in a trophic web. In essence, if there are a variety of trophic links, the same flow of energy and nutrients will be maintained through alternate links when a species becomes rare. However, May's (1973) recent mathematical analysis of models of a large number of interacting populations shows that this relation between increased diversity and stability is not a mathematical truism. He shows that randomly assembled complex systems are in general less stable, and never more stable, than less complex ones. He points out that ecosystems are likely to have evolved to a very small subset of all possible sets and that MacArthur's conclusions, therefore, might still apply in the real world. The definition of stability used, however, is the equilibrium-centered one. What May has shown is that complex systems might fluctuate more than less complex ones. But if there is more than one domain of attraction, then the increased variability could simply move the system from one domain to another (see also Chapter 5 by Prigogine, Section 8). It would be useful to explore the possibility that instability in numbers can result in more diversity of species and in spatial patchiness, and hence in increased resilience.

Measurement

If there is a worthwhile distinction between resilience and stability, it is important that both be measurable. In a theoretical world, such measurements could be developed from the behavior of model systems in

ISBN 0-201-03438-7/0-201-03439-5pbk.

phase space. There are two components that are important: one that concerns the cyclic behavior and its frequency and amplitude, and one that concerns the configuration of forces caused by the positive and negative feedback relations.

To separate the two, we need to imagine first the appearance of a phase space in which there are no such forces operating. This would produce a referent trajectory containing only the cyclic properties of the system. If the forces were operating, departure from this referent trajectory would be a measure of the intensity of the forces. The referent trajectories that would seem to be most useful would be the neutrally stable orbits of Figure 4.1a, for we can arbitrarily imagine these trajectories as moving on a flat plane. At least for more realistic models, parameter values can be discovered that do generate neutrally stable orbits. In the complex predator–prey model of Holling (Holling and Ewing, 1971), if a range of parameters is chosen to explore the effects of different degrees of contagion of attack, the interaction is unstable when attack is random and stable when it is contagious. We have recently shown that there is a critical level of contagion between these extremes that generates neutrally stable orbits. These orbits, then, have a certain frequency and amplitude and the departure of more realistic trajectories from these referent ones should allow the computation of the vector of forces. If these were integrated, a potential field would be represented with peaks and valleys. If the whole potential field were a shallow bowl, the system would be globally stable and all trajectories would spiral to the bottom of the bowl, the equilibrium point. But if, at a minimum, there were a lower extinction threshold for prey, then, in effect, the bowl would have a slice taken out of one side, as suggested in Figure 4.2. Trajectories that initiated far up on the side of the bowl

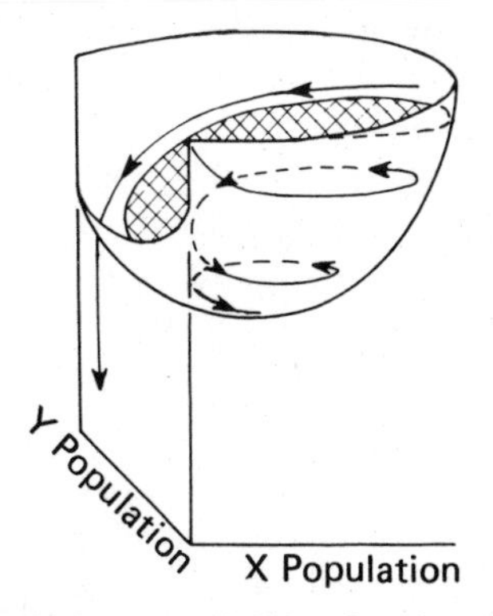

FIGURE 4.2. Diagrammatic representation showing the feedback forces as a potential field upon which trajectories move. The shaded portion is the domain of attraction.

ISBN 0-201-03438-7/0-201-034[illegible] pbk.

would have amplitude that would carry the trajectory over the slice cut out of it. Only those trajectories that just avoided the lowest point of the gap formed by the slice would spiral in to the bowl's bottom. If we termed the bowl the *basin of attraction* (Lewontin, 1969), then the domain of attraction would be determined by both the cyclic behavior and the configuration of forces. It would be confined to a smaller portion of the bottom of the bowl, and one edge would touch the bottom portion of the slice taken out of the basin.

This approach, then, suggests ways to measure relative amounts of resilience and stability. There are two resilience measures: Since resilience is concerned with probabilities of extinction, first, the overall area of the domain of attraction will in part determine whether chance shifts in state variables will move trajectories outside the domain. Second, the height of the lowest point of the basin of attraction (e.g., the bottom of the slice described in the preceding paragraph) above equilibrium will be a measure of how much the forces have to be changed before all trajectories move to extinction of one or more of the state variables.

The measures of stability would be designed in just the opposite way from those that measure resilience. They would be centered on the equilibrium rather than on the boundary of the domain, and could be represented by a frequency distribution of the slopes of the potential field and by the velocity of the neutral orbits around the equilibrium.

But such measures require an immense amount of knowledge of a system, and it is unlikely that we shall often have all that is necessary. Hughes and Gilbert (1968), however, have suggested a promising approach to measuring probabilities of extinction and hence of resilience. They were able to show in a stochastic model that the distribution of surviving population sizes at any given time does not differ significantly from a negative binomial. This, of course, is just a description, but it does provide a way to estimate the very small probability of zero (i.e., of extinction) from the observed mean and variance. The configuration of the potential field and the cyclic behavior will determine the number and form of the domains of attraction, and these will in turn affect the parameter values of the negative binomial or of any other distribution function that seems appropriate. Changes in the zero class of the distribution, that is, in the probability of extinction, will be caused by these parameter values, which can then be viewed as the relative measures of resilience. It will be important to explore this technique first with a number of theoretical models so that the appropriate distributions and their behavior can be identified. It will then be quite feasible, in the field, to sample populations in defined areas, apply the appropriate distribution, and use the parameter values as measures of the degree of resilience.

ISBN 0-201-03438-7/0-201-03439-5pbk.

4. APPLICATION TO THE HUMAN WORLD

Resource Management

The resilience and stability viewpoints of the behavior of ecosystems can yield very different approaches to the management of resources. The stability view emphasizes the equilibrium, the maintenance of a predictable world, and the harvesting of nature's excess production with as little fluctuation as possible. The resilience view emphasizes domains of attraction and the need for persistence. But extinction is not purely a random event; it results from the interaction of random events with those deterministic forces that define the shape, size, and characteristics of the domain of attraction. The very approach, therefore, that assures a stable maximum sustained yield of a renewable resource might so change these deterministic conditions that the resilience is lost or reduced, so that a chance and rare event that previously could be absorbed can trigger a sudden dramatic change and loss of structural integrity of the system.

A management approach based on resilience, on the other hand, would emphasize the need to keep options open, the need to view events in a regional rather than a local context, and the need to emphasize heterogeneity (see also Chapter 10 by Maruyama). Flowing from this would be not the presumption of sufficient knowledge, but the recognition of our ignorance; not the assumption that future events are expected, but that they will be unexpected. The resilience framework can accommodate this shift of perspective, for it does not require a precise capacity to predict the future, but only a qualitative capacity to devise systems that can absorb and accommodate future events in whatever unexpected form they may take.

Societal Implications

It is always tempting, though dangerous, to make analogies—witness the past irrelevancies of the misapplication of Darwinian ideas from biology to anthropology. But we can ask if systems other than natural ecological ones have properties in which the notions of resilience have some application. There are really only two key properties required for such a transfer of concepts.

One property required is that there be more than one equilibrium state, since as soon as that condition prevails, defined stability regions are formed around each equilibrium state and there exists the possibility for sudden movements from one qualitative kind of behavior to

ISBN 0-201-03438-7/0-201-03439-5pbk.

another. It must, I would argue, be almost a truism that physical, human behavioral, institutional, and societal systems must always have multiple stable regions. Individual human responses to given situations that trigger rage versus fear, or flight versus approach, are examples of such qualitatively different responses to the same stimulus. In a more extreme form the classic shifts in some mentally disturbed individuals from manic to depressive states is a further example; and what is true of individuals is equally true of human institutions. The very fact that human institutions respond to circumstances of unexpected events by labeling them crises is an indication that a limited range of disturbances can be absorbed by an institutional system and if those limits are exceeded, the normal corrective responses no longer work.

It is tempting, too, to argue that our present-day industrial societies have moved beyond one stability region into another. Since the mid or late 1960's a variety of societal indicators—of crime, of land costs, of inflation—have all shown rapid increases that are at least suggestive of movements into a different stability configuration. It could be argued that the emphasis on efficiency and optimal solutions in our society, while ignoring limitations in resource and human behavioral attributes, has led to a transient improvement in the quality of life at the expense of the gradual shrinkage of a stability domain within which such conditions can persist. The stability region, therefore, instead of becoming a comfortably wide pit, becomes so contracted as to poise our society delicately on a mountain top, where the slightest disturbance can topple us to a dramatically new qualitative condition. But that is the rankest of speculation. We can turn with somewhat more substance to anthropological examples which provide rather more convincing evidence.

As but one example, Rappoport (1967) presents an interesting analysis of the role of ritual in the regulation of environmental relations among a New Guinea society. In its simplest form this society obtains its food from the surrounding forest, market gardens, and pigs. But there is a taboo on eating pigs except on special ceremonial occasions. These ceremonial occasions are triggered when the social temperature—conflict—reaches a critical point in the village. At this point a ceremony of propitiation to the gods occurs in which the key element is the exclusive consumption of pigs. But by and large the reason the conflict occurs is because the high pig populations begin to interfere with the market gardens. Neighbor becomes irritated with neighbor and, magically, after the feast of propitiation the problems disappear.

This is in no sense an example of an optimal food production system that produces low degrees of fluctuation. In fact, quite the opposite. It is as if a ritualized mechanism had developed to assure a wide

ISBN 0-201-03438-7/0-201-03439-5pbk.

degree of fluctuation within that society so that the stability boundaries of their system could be continually monitored. A key feature of this ritual is that not only is the fluctuation assured but, more significantly, strong mechanisms are developed to turn the society away from a stability boundary as the signals are detected. Rather than minimizing the probability of difficulty, this society seems to have a designed method of generating detectable but controllable "failures." They occur frequently enough to prevent stability regions from contracting by maintaining flexibility of institutional response. That really is not too different from China's cultural revolution approach.

In these few examples, therefore, it seems reasonable to presume the existence of multiple stability regions in individuals, institutions, and societies. But there is another key property that is important in relation to resilience, and that property recognizes that stability characteristics of a system are not static. Given a particular system, the parameters of that system can evolve in time so that stability boundaries can contract or expand. This occurs through evolutionary forces. It becomes very evident in ecological systems where the stability regions are dictated typically by events at low densities of organisms—essentially thresholds of extinction. Those thresholds of extinction have evolved as a consequence of a balanced natural selection. If management procedures are set in motion to limit the fluctuations of the organisms, then the balance of natural selection can shift and the extinction thresholds thereby can begin to move. The consequence, in such examples, is a shrinking of stability regions—a shrinking that is not at all evident until an unexpected perturbation, that previously could be absorbed, no longer can be.

Fail-safe versus Safe-fail Strategies

If these resilience notions, and the possibilities of sharp jumps in behavior as a consequence of shrinking domains of stability, are at all relevant to societal systems, then we have the opportunity to begin to explore alternate strategies for societal design. I started this chapter by arguing that there were two views of the world, one of which emphasized the *optimal* and the *quantitative* and the other of which emphasized *persistence* or *resilience* and *qualitative* states. These two different views in their simplest expression could represent two extreme strategies for design. The first would lead to the design of highly optimal systems in which fluctuations were minimized and explicit efforts were made to minimize the probability of failure—in short, a *fail-safe strategy.* There is absolutely nothing wrong with that if, accompanying that strategy, is

ISBN 0-201-03438-7/0-201-03439-5pbk.

an explicit ability to maintain a stability region of a size large enough to contain any unexpected perturbation the system might receive. That, it could be argued, might be appropriate for those simple systems that, like some physical technological ones, are extremely well known—in which, for example, the properties of metal fatigue in molecular structure are so completely grasped that a physical item can be designed to give complete assurance that its stability region will not contract. In such an example, the goal of safety would be achieved by maximizing the distance from the static and invariant stability boundary.

But rarely do we have that kind of knowledge, and certainly not in societal systems. In societal systems, or for that matter ecological ones, we have only the roughest idea where stability boundaries exist and their cause, and we certainly have no confidence at all that those existing stability boundaries are invariant in time. If anything, we could argue that it was almost inevitable that an attempt to reduce fluctuations would, at the same time, cause a shrinkage of the stability region through the action of cultural or natural selection. That possibility requires not a fail-safe strategy but one that is *safe-fail*—that is, that optimizes a cost of failure and even assures that there are periodic "minifailures" to prevent evolution of inflexibility.

The specific strategy adopted for any situation in reality should be a mix of these two extremes, the mix determined by our degree of knowledge. A crude existing example is the microsociety of a cruise ship. In itself that society and the infrastructure that supports it provides a clear example of the two contrasting strategies of design. The ship itself is designed with an optimal, equilibrium-centered strategy in the fore. The best of knowledge of ship design and stability is incorporated within the structure of the ship so as to minimize the probability of failure. But through bitter experience, society has learned that the probability of failure can never be reduced to zero—ships do founder, mutinies do occur. So through experience there has evolved a tradition of attempting, in essence, to mimic disasters—to mimic the fluctuations that are so necessary to prevent cultural evolutionary forces from acting so as to contract stability domains. In the example of the cruise ship those mimics of disaster are the lifeboat drills and the fire drills. The more often they are performed and with the greatest realism, the more likely it is that the microsociety of the cruise ship can adapt to and absorb unexpected events.

In any real sense, therefore, either one of those extreme strategies is probably inappropriate. As argued earlier, if the system is totally known, so that the stability regions can be part of the design effort and can be fixed, then a simple equilibrium-centered optimal strategy is ideal, particularly if the distance to the static stability boundary is max-

ISBN 0-201-03438-7/0-201-03439-5pbk.

imized. But if the world is partially known, as surely most of our world is, then the design goal should be quite different. In that instance the attempt should be to carefully optimize key requirements but maintain the fluctuations of the system as close to the unmodified one as possible. In that way, in a sense, the variables themselves are used to continually monitor the stability boundary in order to maintain the size of that region. As noted in the example of the cruise ship, those fluctuations need not exactly occur but can be mimicked.

Finally, in a totally unknown, or largely unknown, hypothetical world, the only reasonable design criterion follows the progression described in the foregoing to its logical conclusion. That is, one should attempt to *keep as close to a stability boundary as possible*, for it is only at that point that one can easily shift to one stability condition or another, as information is gained as to the state desired. The farther one moves away from a stability region in an unknown world, the more difficult it is to move away from that state if evidence accumulates to suggest one *should* move away from it.

Our existing efforts to design a new society tend to erect new societal goals for old ones, and presume we move invariably toward them. This is as fragile and as doomed to crises as our present society is, and that is true even if those new goals emphasize conservation and ecological sensitivity. The longer institutions and attitudes aim toward that new goal, the more resistant they would be to moving away from it if the need for such movement emerged. That leads therefore to the suggestion that with such unknowns, the place where options are kept most open is near stability boundaries. It leads therefore to the conclusion that increasingly we must learn to *live with disturbance, live with variability, and live with uncertainties.* Those are the ingredients for persistence.

REFERENCES

Baskerville, G. L. (1971). *The Fir–Spruce–Birch Forest and the Budworm.* Report, Forestry Service, Canada Dept. Environ., Fredericton, N. B.

Elton, C. S. (1958). *The Ecology of Invasions by Animals and Plants.* London: Methuen.

Griffiths, K. J., and Holling, C. S. (1969). "A Competition Submodel for Parasites and Predators," *Can. Entomol.*, **101**, 785–818.

Holling, C. S. (1961). "Principles of Insect Predation," *Ann. Rev. Entomol.*, **6**, 163–82.

ISBN 0-201-03438-7/0-201-03439-5pbk.

Holling, C. S. (1965). "The Functional Response of Predators to Prey Density and Its Role in Mimicry and Population Regulation," *Mem. Entomol. Soc. Can.*, **45**, 1-60.

Holling, C. S. (1966). "The Functional Response of Invertebrate Predators to Prey Density," *Mem. Entomol. Soc. Can.*, 48, 1-86.

Holling, C. S., and Ewing, S. (1971). "Blind Man's Buff: Exploring the Response Space Generated by Realistic Ecological Simulation Models," *Proc. Intern. Symp. Stat. Ecol.*, New Haven, Conn.: Yale Univ. Press.

Hughes, R. D., and Gilbert, N. (1968). "A Model of an Aphid Population—A General Statement," *J. Animal Ecol.*, **40**, 525-534.

Lewontin, R. C. (1969). "The Meaning of Stability," in *Diversity and Stability of Ecological Systems, Brookhaven Symp. Biol.*, **22**, 13-24.

MacArthur, R. (1955). "Fluctuations of Animal Populations and a Measure of Community Stability," *Ecology*, **36**, 533-536.

May, R. M. (1973). *Model Ecosystems.* Princeton, N.J.: Princeton Univ. Press.

Minorsky, N. (1962). *Nonlinear Oscillations.* Princeton, N.J.: Van Nostrand.

Morris, R. F. (1963). "The Dynamics of Epidemic Spruce Budworm Populations," *Mem. Entomol. Soc. Can.*, 31, 1-332.

Nicholson, A. J., and Bailey, V. A. (1935). "The Balance of Animal Populations—Part I," *Proc. Zool. Soc. London*, **1935**, 551-598.

Rappoport, R. A. (1967). "Ritual Regulation of Environmental Relations among a New Guinea People," *Ethnology*, **6**, 17-30.

Rosenzweig, M. L. (1971). "Paradox of Enrichment: Destabilization of Exploitation Ecosystems in Ecological Time," *Science*, **171**, 385-387.

Rosenzweig, M. L. (1972). "Stability of Enriched Aquatic Ecosystems," *Science*, **175**, 564-565.

Rosenzweig, M. L., and MacArthur, R. H. (1963). "Graphical Representation and Stability Condition of Predator-Prey Interactions," *Amer. Natur.*, **97**, 209-223.

Slobodkin, L. B. (1964). "The Strategy of Evolution," *Amer. Sci.*, **52**, 342-357.

Smith, S. H. (1968). "Species Succession and Fishery Exploitation in the Great Lakes," *J. Fish. Res. Bd. Can.*, **25**, 667-693.

Walters, C. J. (1971). "Systems Ecology: The Systems Approach and Mathematical Models in Ecology," in *Fundamentals of Ecology* (E. P. Odum, ed.), 3d ed. Philadelphia: Saunders.

Wangersky, P. J., and Cunningham, W. J. (1957). "Time Lag in Prey-Predator Population Models," *Ecology*, **38**, 136-139.

Wellington, W. G. (1952). "Air Mass Climatology of Ontario North of Lake Huron and Lake Superior before Outbreaks of the Spruce Budworm and the Forest Tree Caterpillar," *Can. J. Zool.*, **30**, 114-127.

Wellington, W. G. (1964). "Qualitative Changes in Populations in Unstable Environments," *Can. Entomol.*, **96**, 436-451.

Wellington, W. G. (1965). "The Use of Cloud Patterns to Outline Areas with Different Climates during Population Studies," *Can. Entomol.*, **97**, 617-631.

ISBN 0-201-03438-7/0-201-03439-5pbk.

Order through Fluctuation: Self-Organization and Social System

Ilya Prigogine

SUMMARY

In most of the phenomena studied in classical physics, fluctuations play only a minor role. This is the case in the whole domain of classical equilibrium thermodynamics based on Boltzmann's ordering principle. On the other hand, the study of nonlinear systems under conditions far from equilibrium leads to new situations in which fluctuations play a central role. It is the fluctuations that can force the system to leave a given macroscopic state and lead it on to a new state which has a different spatiotemporal structure.

The study of dissipative structures illustrates precisely this type of behavior. Contrary to equilibrium structures, dissipative structures occur at a sufficient distance from thermodynamic equilibrium when the system is described by equations containing an appropriate feedback.

The thermodynamic theory of dissipative structures will be briefly sketched. Recent results show the importance of these structures in numerous problems of chemistry and biochemistry. Dissipative structures lead to a whole spectrum of characteristic dimensions linked to chemical reactions and transport phenomena. Furthermore, examples show that the formation of these structures can be accompanied by symmetry breaking and the appearance of new forms and shapes. Examples will also be presented from the domains of biological aggregation and animal societies.

Dissipative structures can be considered as giant fluctuations; therefore, their evolution over time contains an essential stochastic element. A stochastic equation introduced recently by Nicolis and the author permits the discussion of both the deterministic and the stochastic aspects of these phenomena. In particular, such equations permit the study of the *nucleation* of a new dissipative structure. They also make possible a discussion of the relation between the three levels of description represented by the scheme

ISBN 0-201-03438-7/0-201-03439-5pbk.

Erich Jantsch and Conrad H. Waddington (eds.), *Evolution and Consciousness: Human Systems in Transition.*

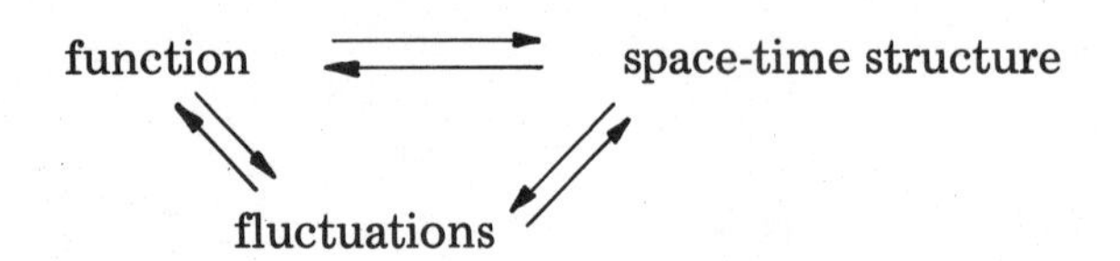

Determinism and fluctuations play a complementary role in our description. It is interesting to apply the formalism to a description of the structure of societies in which the dialectic between "mass and minority" (F. Perroux, 1964) plays an essential role.

1. INTRODUCTION

Both physics and chemistry are at present passing through a period of rapid growth resulting in particular from the integration of the concept of structure into the framework of theoretical physics. In addition, a more precise interpretation of the notions of irreversibility and process has now been given. As a result, a renewed dialogue between researchers in the physical sciences and those interested in the human sciences has become possible.

In the context of modern science, the possibility of such a dialogue was raised right from the first formulation of modern science by Newton and Galileo. It is fascinating to read the chapter "Generalization of the Newtonian Paradigm for Natural Sciences and Human Sciences" in the outstanding monograph by Gusdorf (1971). But this dialogue encountered insurmountable difficulties. Too big a gap separates rational mechanics and the study of simple motions from the specific problems posed by biology or history. This gap is particularly marked in the concept of time. Rational mechanics knows only reversible time, whereas the direction of time plays a fundamental role in the biological and human sciences. It is true that in the nineteenth century, the problem of the direction of time was incorporated in physics through the second law of thermodynamics. Still, the contrast between the idea of evolution in physics and that in biology or sociology is striking. In physics, the increase of entropy expressed by the second law of thermodynamics shows a tendency toward a progressive "disorganization" of the system. On the other hand, biological or social evolution is accompanied by progressive structuration such as that introduced by the division of labor in the history of human societies.

Despite such difficulties, numerous references to physics may be found in the works of specialists in the human sciences. One of the chapters in the discussions between Lévi-Strauss and G. Charbonnier (Charbonnier, 1969) is entitled "Clocks and Steam Engines" and the treatise *Le Système Social* by Henri Janne (1963) starts from the con-

ISBN 0-201-03438-7/0-201-03439-5pbk.

cept of social "force." Thus, the vocabulary of classical physics has spread to the human sciences, at least in a metaphorical sense.

Auguste Comte (see Aron, 1967, p. 82) distinguished between "analytic" and "synthetic" sciences. To the latter belong biology and sociology. In biology, it is impossible to explain an organ, or a function, without referring to the whole living being, and this remark, suitably transposed, applies equally to sociology. The consideration of the whole is essential in both cases. Recently, the study of dissipative structures (see, in particular, Glansdorff and Prigogine, 1971; Prigogine, 1972; Prigogine et al., 1972) has brought the study of such "totalities" within the framework of an extended thermodynamics. We shall return to this concept of dissipative structures in Section 2.

Let us simply recall that classical thermodynamics permits the interpretation of *equilibrium* structures, those that appear, for example, in an isolated system after a sufficiently long time. A crystal is a typical example of such an equilibrium structure. The formation of such structures is dominated by Boltzmann's ordering principle (which gives the population of the different dynamic states in a system at thermodynamic equilibrium). The situation changes radically when instead of a closed system one considers an open system exchanging matter and energy with the outside environment. In this case, providing the reservoirs of energy and matter are sufficiently large to remain unchanged, the system can tend to a constant regime corresponding to a nonequilibrium stationary state. While an isolated system at equilibrium is associated with "equilibrium" structures such as the above-mentioned crystal, an open system "out of equilibrium" may be associated with dissipative structures (when specific constraints, to be discussed later, are satisfied). Then, Boltzmann's ordering principle is no longer applicable. Dissipative structures are associated with an entirely different ordering principle, which may be called *order through fluctuation*. In fact, such structures arise from the amplification of fluctuations resulting from an instability of the "thermodynamic branch."

As we shall see in Section 2, dissipative structures present precisely the global aspect, the aspect of totality, which Comte ascribed to the object of the synthetic sciences.

In order to be able to take form, a dissipative structure requires a nonlinear mechanism to function. It is this mechanism which is responsible for the amplification mechanism of the fluctuation. Dissipative structures thus form a bridge between *function* and *structure*. One may even consider such sociologists as Comte, Durkheim, or Spencer as forerunners of the concept of dissipative structures. In his treatise on the division of social labor, Durkheim (1973, p. 244) writes, for example:

> The division of labor progresses the further the more individuals there are who are in sufficiently close contact with each other to be able to act and react upon each other. If we agree to call this close-

ISBN 0-201-03438-7/0-201-03439-5pbk.

ness and the active transactions resulting from it, dynamic or moral density, then we may say that the progress made by the division of labor is due directly to the moral or dynamic density of society.

The distinction between the ordering principle of Boltzmann and that by fluctuations implies a fundamental difference in the role of the fluctuations. In the domain in which Boltzmann's principle is applicable, the fluctuations play a subordinate role. Consider, for example, a volume V within which we have N particles. Let us study a small volume ΔV within V (see Figure 5.1). We expect that the average number of particles $\bar{n}$ in ΔV will be

$$\frac{\bar{n}}{N} = \frac{\Delta V}{V} \tag{1.1}$$

Of course there will be fluctuations. The fluctuations permit us, however, to specify the mean result (1.1) but can be neglected in many situations (they are of the order of $\bar{n}^{1/2}$). On the contrary, in the formation of dissipative structures, it is the fluctuations that drive the system to a new average state. In other words, instead of being simply a corrective element, the fluctuations become the *essential* element in the dynamics of such systems. Here again, the analogy with characteristic situations in biology and sociology is striking.

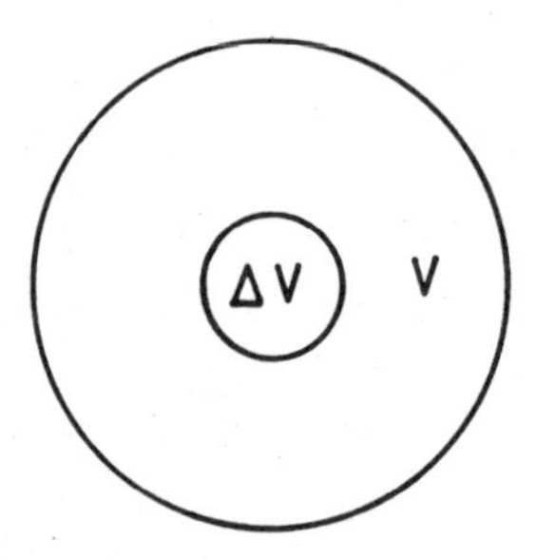

FIGURE 5.1.

Many authors have indeed stressed the double role of "chance and necessity" in the human sciences and it is precisely these two elements which play a role in the phenomena dominated by the principle of order through fluctuation.

In the present chapter, we shall first illustrate the concepts of dissipative structure and order through fluctuation by presenting some simple examples borrowed from physical chemistry and from the study of social insects. Subsequently, we shall consider the statistical description of such systems and shall briefly recall the nonlinear stochastic equations

ISBN 0-201-03438-7/0-201-03439-5pbk.

introduced recently by Nicolis and the author (see Nicolis and Prigogine 1971; Nicolis et al., 1974; Prigogine et al., 1974). A more detailed discussion of self-organization in nonequilibrium systems may be found in the monograph by Nicolis and Prigogine (1976).

2. THERMODYNAMICS AND DISSIPATIVE STRUCTURES

If we consider first an isolated system, the second law of thermodynamics tells us that the entropy production within the system, due to irreversible phenomena, is positive (Clausius, 1857):

$$dS = d_{\mathrm{i}}S > 0 \tag{2.1a}$$

if the system undergoes an irreversible transformation, and

$$dS = d_{\mathrm{i}}S = 0 \tag{2.1b}$$

if the system is at equilibrium. When there are exchanges with the outside world, (2.1a, b) must be completed by a term of entropy flux $d_{\mathrm{e}}S$

$$dS = d_{\mathrm{e}}S + d_{\mathrm{i}}S \tag{2.2}$$

Only $d_{\mathrm{i}}S$ has a well-defined sign. Identifying entropy with disorder (Boltzmann, 1972), we see that an isolated system can only evolve toward greater disorder. For an open system, however, the competition between $d_{\mathrm{e}}S$ and $d_{\mathrm{i}}S$ permits the system, subject to certain conditions which are made precise later, to adopt a new, *structured* form.

The entropy production can be expressed simply in terms of thermodynamic "forces" X_i and "rates" of irreversible phenomena J_i (Prigogine, 1967). For example, X_i may be temperature gradients, chemical "affinities," or the like. The corresponding "rates" are then heat flux and chemical reaction rate. We have

$$\frac{d_{\mathrm{i}}S}{dt} = \sum_i J_i X_i \tag{2.3}$$

At thermodynamic equilibrium one has simultaneously

$$J_i = 0 \qquad X_i = 0 \tag{2.4}$$

ISBN 0-201-03438-7/0-201-03439-5pbk.

whereas around equilibrium, in the so-called linear domain of thermodynamics, we have a linear relation between fluxes and forces:

$$J_i = \sum_j L_{ij} X_j \tag{2.5}$$

Onsager (1931) has shown that the coefficients L_{ij} form a symmetric matrix (Onsager's "reciprocity" relations).

In addition, the author has shown that in this linear domain the entropy production can only decrease in time and will attain its minimum value for the stationary state compatible with the imposed conditions (Prigogine, 1967). The linear domain (2.5) thus extends in this way the equilibrium behavior (2.4). To obtain a new structuration and a behavior which is radically different from that of equilibrium, we must go beyond the domain of linear thermodynamics.

There exists in all systems a thermodynamic threshold above which the system can exhibit "self-organization" if it is the seat of appropriate irreversible phenomena. Let us consider more closely an example of nonlinear chemical networks which can lead to self-organization.

3. AN EXAMPLE OF A DISSIPATIVE STRUCTURE: THE TRIMOLECULAR MODEL

The behavior and nature of a dissipative structure that can appear beyond a certain critical distance from equilibrium may be multiple: temporal organization (limit cycle), stationary inhomogeneous structure, spatiotemporal organization in a wave form, and localized structures.

A model chemical reaction, somewhat unrealistic from the experimental point of view, but offering both great wealth of behavior and facility of analysis, has been developed by the Brussels group (Nicolis and Auchmuty, 1974; Lefever 1968; Herschkowitz-Kaufman, 1973).

$$A \rightleftharpoons X \tag{3.1a}$$

$$B + X \rightleftharpoons Y + D \tag{3.1b}$$

$$2X + Y \rightleftharpoons 3X \tag{3.1c}$$

$$X \rightleftharpoons E \tag{3.1d}$$

The trimolecular step (3.1c) can be considered as the result of several very rapid bimolecular steps. The concentrations of the products A, B, D, E are maintained constant and are the constraints that permit the system to be driven away from equilibrium.

ISBN 0-201-03438-7/0-201-03439-5pbk.

If we set the kinetic constants equal to 1 for the sake of simplified writing, the equations for the evolution of the concentrations X, Y in a unidimensional environment are as follows:

$$\frac{\partial X}{\partial t} = A + X^2 Y - (B + 1)\, X + D_X \frac{\partial^2 X}{\partial r^2} \tag{3.2a}$$

$$\frac{\partial Y}{\partial t} = BX - X^2 Y + D_Y \frac{\partial^2 X}{\partial r^2} \tag{3.2b}$$

where $0 \leqslant r \leqslant \ell$.

These nonlinear partial differential equations translate the effect of the chemical reactions and of the diffusion. D_X and D_Y are the diffusion coefficients (Fick's law).

The system can assume a single homogeneous stationary state

$$X_{st} = A, \qquad Y_{st} = B/A \tag{3.3}$$

The boundary conditions are fixed in the stationary state

$$X(0) = X(\ell) = A, \qquad Y(0) = Y(\ell) = B/A \tag{3.4}$$

The study of the system of linearized equations around this stationary state shows that the latter may become unstable if A, B, D_X, and D_Y satisfy certain relations.

Different types of behavior are possible (see also Chapter 4 by Holling):

1. In the case for which D_X and D_Y are large, the system may be considered as homogeneous. It may present an instability of the *limit cycle type*: the concentrations X and Y oscillate around the stationary state. The evolution of the system in phase space is shown in Figure 5.2. Thus, S marks the point corresponding to the unstable stationary state (X_S, Y_S). Whatever its initial state, the system tends, in the course of time, toward a single well-defined periodic solution whose characteristics (period and amplitude) are imposed by the differential equation. *These oscillations are stable* and very different from the type of periodic behavior presented by a *Lotka–Volterra system*. In the latter case, the oscillations are not stable with respect to the fluctuations (for a detailed analysis see Glansdorff and Prigogine, 1971). There exists an infinity of orbits surrounding the stationary state. Analysis shows that there is no "restoring" force for any particular orbit, and any little perturbation can change the system from one orbit to another. This behavior differs radically from that of a limit cycle when only one stable periodic solution exists.

ISBN 0-201-03438-7/0-201-03439-5pbk.

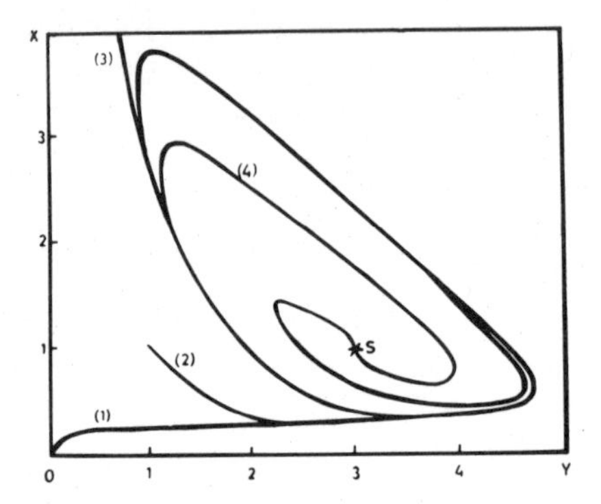

FIGURE 5.2. Trajectories tending asymptotically to the limit cycle, obtained by numerical integration for different initial states (with $A = 1, B = 3$): (1) $X = Y = 0$; (2) $X = Y = 1$; (3) $X = 10, Y = 0$; (4) $X = 1, Y = 3$. (See Lefever, 1968.)

When D_X and D_Y are not large enough, the system acquires a spatiotemporal regime corresponding to the propagation of concentration waves or of stationary chemical waves (see Herschkowitz-Kaufman and Nicolis, 1972; Auchmuty and Nicolis, 1975, 1976; and Nicolis and Prigogine, 1974). One observes (Figure 5.3) the existence of a stable periodic regime which corresponds to the alternation of a rapid phase of chemical waves propagating toward the center and a slow phase tending to homogenize the system.[1]

2. The system can evolve toward a new stable stationary state but where X and Y are distributed inhomogeneously. Figure 5.4 shows an example of an inhomogeneous distribution of X as a function of space. Among the properties of the solutions beyond instability it is interesting to note the possibility of the spontaneous formation of polarity in a system under the effect of a perturbation (see Babloyantz and Hiernaux, 1975). This observation is particularly interesting for explaining the appearance of inhomogeneities during the development of an embryo from an initially unfertilized, homogeneous egg. Figure 5.5 represents the inhomogeneous distribution of X in polar coordinates obtained after perturbing the stationary homogeneous state. According to the exact location of the initial perturbation, the gradient of X will be oriented in one sense or the other.

The existence of *localized structures* has been demonstrated with this model. When the initial product A diffuses through the system, then the equation for A must be added to (3.2):

$$\frac{\partial A}{\partial t} = -A + D_A \frac{\partial^2 A}{\partial r^2} \qquad (3.5)$$

[1] Nicolis and Auchmuty (1974) have shown that chemical waves have to be considered as a superposition of stationary waves. There exists in general no well-defined velocity of propagation.

ISBN 0-201-03438-7/0-201-03439-5pbk.

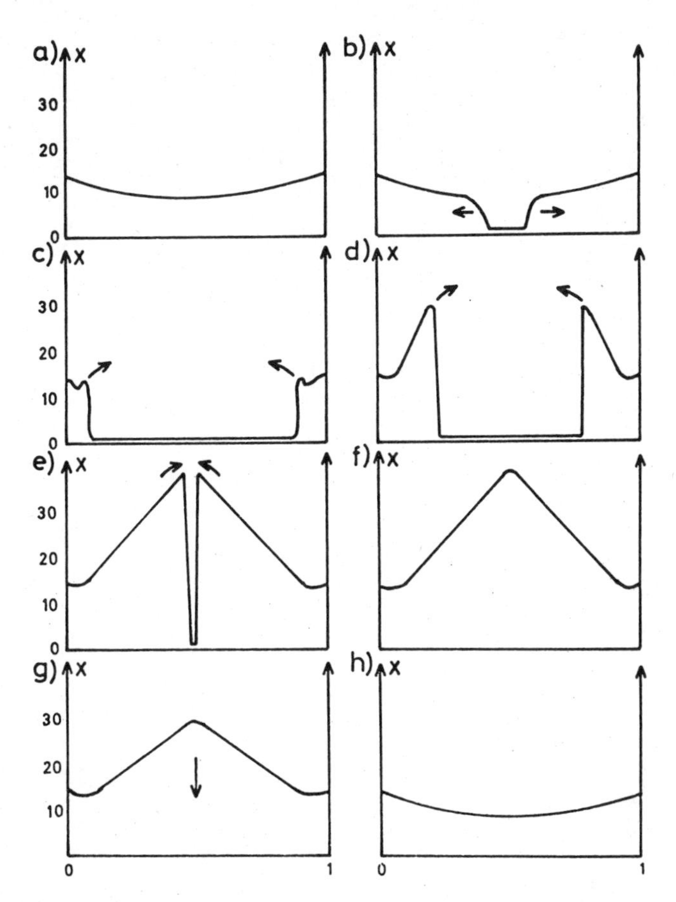

FIGURE 5.3. Characteristic steps in the evolution of the spatial distribution of the intermediate X. $D_X = 0.00105$, $D_Y = 0.00066$, $D_A = 0.195$; $B = 77$; $X(0) = X(1) = 14$. (See Herschkowitz-Kaufman and Nicolis, 1972.)

Figure 5.6 shows the stationary dissipative structure engendered under these conditions beyond a critical point of instability. One notices that this time the spatial organization is limited to a small region of the possible domain. Outside these frontiers, the concentration distributions correspond to the thermodynamic solution.

The dissipative structure can appear as a "totality" with its dimensions imposed by its own mechanism. Conversely, the dimensions of the system play an essential role in the formation of dissipative structures. A sufficiently small system will always be dominated by the boundary conditions (Nazarea, 1974). In order for the "nonlinearities" to be able to lead to a choice between various solutions, it is necessary to go beyond some critical spatial dimensions (Hanson, 1974). It is only then that the system aquires some autonomy with respect to the outside world.

ISBN 0-201-03438-7/0-201-03439-5pbk.

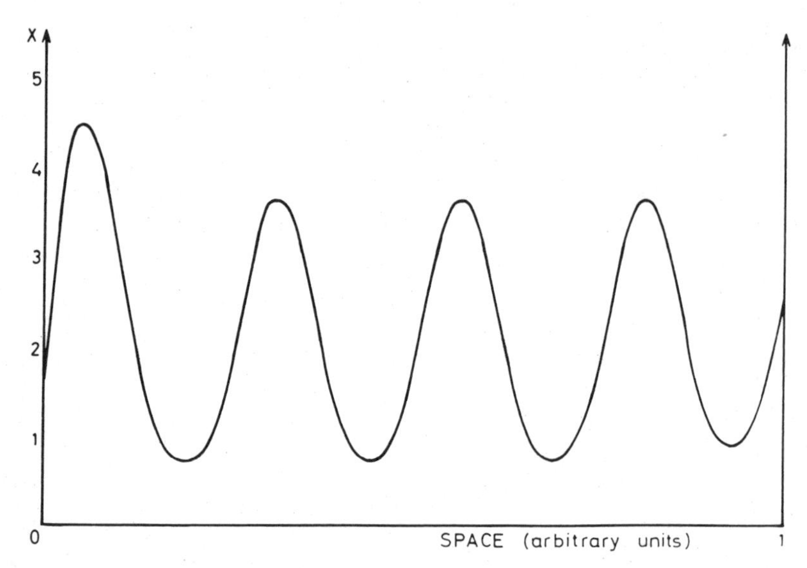

FIGURE 5.4. A dissipative structure of X when $A = 2$; $B = 4.6$; $D_X = 0.0016$; $D_Y = 0.0080$. (See Herschkowitz-Kaufman, 1973.)

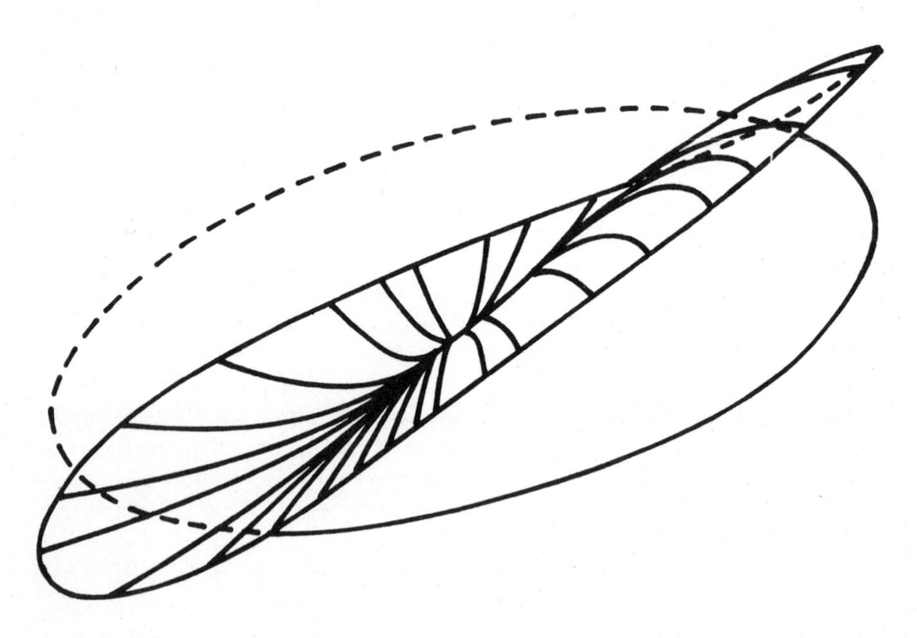

FIGURE 5.5. A two-dimensional spatial dissipative structure. The flux is taken as zero at the boundaries. Diameter, 0.2; $D_X = 0.00325$; $D_Y = 0.0162$; $B = 4.6$. (See Erneux and Herschkowitz-Kaufman, 1975.)

ISBN 0-201-03438-7/0-201-03439-5pbk.

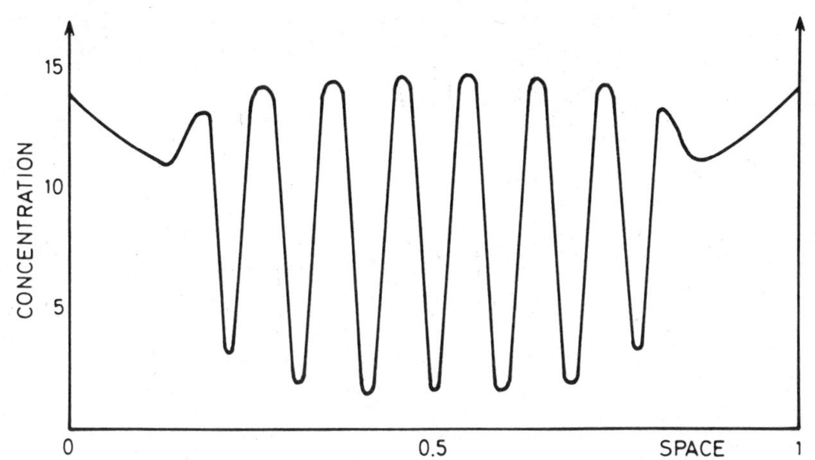

FIGURE 5.6. A stationary solution of Eqs. (3.2) and (3.5) beyond the critical point of instability for the boundary conditions of X, Y, and A given for $B = 26$. The reduced values adopted were $D_A = 0.197$; $D_X = 0.00105$; $D_Y = 0.00526$. (See Herschkowitz-Kaufman and Nicolis, 1972.)

Experimental observations have confirmed the existence of these remarkable properties associated with dissipative structures, be they biochemical, organic, or mineral systems (see Faraday Symposium 9, 1974, especially Boiteux and Hess, 1974).

4. EVOLUTION—STRUCTURAL STABILITY

Until now we have examined the problem of the organization of physicochemical systems without discussing the possibility of a change in the laws of chemical kinetics. Indeed we have supposed that the mechanism (3.1) remains the same on the thermodynamic branch or in the structured domain. We shall now examine the problem of stability when fluctuations may modify the kinetics, for example, through the formation of new substances. A similar situation exists in biological evolution, which describes the appearance of new species. Also, in the social domain, behavioral fluctuations can lead to a modification of social structures.

All these problems are related to the theory of *structural stability* in which the stability of the equations is studied with respect to small perturbations leading to "new" forms of kinetics. If a new substance appears after such a perturbation, its concentration can either diminish (and the system returns to its initial behavior) or be amplified, which then results in the appearance of a new macroscopic mechanism.

ISBN 0-201-03438-7/0-201-03439-5pbk.

Let us briefly clarify these notions (see Prigogine et al., 1972) before discussing specific examples. First let us consider a sequence of polymerization reactions

$$X_1 + X_1 \rightarrow X_2$$
$$X_2 + X_1 \rightarrow X_3$$
$$X_3 + X_1 \rightarrow X_4$$
$$\vdots \qquad \vdots$$

To these steps we may add terms representing different types of catalytic effects. For example, the concentration of X_2 may, in addition, be favored by the autocatalytic process

$$X_1 + X_1 + X_2 \rightarrow 2X_2$$

The system contains N chemical substances $\{X_i\}$, $i = 1, 2, \ldots, N$. The evolution in time is given by

$$\dot{X}_i = F_i^e(X_1, X_2, \ldots, X_N) + \bar{F}_i(X_1, X_2, \ldots, X_N) \qquad (4.1)$$

where F_i^e represents the flux of matter, and $\bar{F}_i$ describes the reactions taking place inside the system. The F_i^e are maintained constant and uniform in the system.

The stability of Eq. (4.1) can be studied by linearizing around the stationary state. The concentrations of the different species are developed around this state

$$X = X_S + \tilde{X} e^{\omega t}$$

where X_S is the value at the stationary state and $\tilde{X}$ is the initial perturbation. Substituting X into the system of equations, we obtain a secular equation for ω whose order is equal to the number of variables:

$$a_0\omega^N + a_1\omega^{N-1} + \ldots + a_N = 0 \qquad (4.2)$$

The solution of this equation tells us how the system will evolve: if ω has an imaginary part, the system will oscillate; if the real part of ω, ω_r, is positive, the perturbation will be amplified; and if ω_r is negative, it will regress.

ISBN 0-201-03438-7/0-201-03439-5pbk.

Let us suppose that the stationary solution of (4.1) is stable. Then all the solutions of the secular equation must have a negative real part. The fluctuations only modify the concentrations of the different species in the system.

However, another type of fluctuation can occur, one leading not to an alteration in the concentration of the existing X_i, but to the formation of new substances $Y_i (i = 1, \ldots, m)$. In place of (4.1), we have therefore a new system with m supplementary differential equations for the new substances Y_i. The order of the secular determinant giving the ω_i becomes $N + m$. If we take $m = 1$, the secular equation of degree $N + 1$ has the form of either

$$\epsilon\omega^{N+1} + a'_0(\epsilon)\omega^N + \ldots + a'_N(\epsilon) = 0 \tag{4.3}$$

or

$$a'_0(\epsilon)\omega^{N+1} + a'_1(\epsilon)\omega^N + \ldots + \epsilon a'_{N+1}(\epsilon) = 0 \tag{4.4}$$

with

$$\lim_{\epsilon \to 0} a'_j(\epsilon) = a_j, \qquad j = 0, 1, \ldots, N \tag{4.5}$$

For $\epsilon \neq 0$, but small, the $N + 1$ roots of the new system $\{X_i\} + Y$ must have N roots almost equal to those of the equation without the new substance Y (Andronov et al., 1966). In particular, the real parts must have the same sign in both cases. The stability of the stationary state can only be compromised by the $(N + 1)$th root. Let us indicate the form of the supplementary root ω_{N+1}. From (4.3) it would follow

$$\omega_{N+1} \simeq -a'_0/\epsilon \tag{4.6}$$

while from (4.4) we would have

$$\omega_{N+1} \simeq -\epsilon a'_{N+1}/a'_N \tag{4.7}$$

If $\epsilon > 0$, ω_{N+1} could have a positive real part according to the sign of a'_0, a'_N, and a'_{N+1}. We arrive, therefore, at the important conclusion that the addition of a new substance Y can destroy the previously existing stability of the system. The characteristic of the problem of structural stability calls for the study of stability with respect to the appearance of substances that were initially absent from the original scheme (4.1).

ISBN 0-201-03438-7/0-201-03439-5pbk.

Let us now examine more closely the respective situations to which the secular equations (4.3) and (4.4) correspond. We can easily verify that the equation of evolution of Y, corresponding to a secular equation of the form of (4.3), must be of the form

$$\epsilon \dot{Y} = G(\{X_i\}, Y, \epsilon) \tag{4.8}$$

Furthermore, the presence of Y modifies the evolution equations of the $\{X_i\}$, which become, instead of (4.1),

$$\dot{X}_i = F_i^e + F_i(\{X_i\}, Y, \epsilon) \tag{4.9}$$

with the obvious conditions

$$G(\{X_i\}, Y, 0) = 0 \tag{4.10}$$

$$F_i(\{X_i\}, Y, 0) \equiv \overline{F}_i(\{X_i\}) \tag{4.11}$$

The equations for Y do not contain a flux term, because this substance is produced by the $\{X_i\}$. On the contrary, the secular equation (4.4) imposes kinetic equations of the form

$$\dot{Y} = G_1(\{X_i\}) + \epsilon G_2(\{X_i\}, Y, \epsilon) \tag{4.12}$$

$$\dot{X}_i = F_i^e + F_{1i}(\{X_i\}) + \epsilon F_{2i}(\{X_i\}, Y, \epsilon) \tag{4.13}$$

The essential difference between (4.8) and (4.12) is that in (4.8) a new, shorter time scale appears because of the fluctuation. Indeed, the rate of formation of Y (4.8) is proportional to $1/\epsilon$, the parameter ϵ being considered small. The fluctuation leads to an "acceleration" of the polymerization process. When there are different mechanisms competing in the system, it is clear why it should be the structural instability of type (4.8) and not (4.12) that plays an essential role (Eigen, 1971). These concepts have been applied to the interpretation of prebiotic evolution. Errors of transcription in the chemical kinetics lead to the production of new substances possessing a greater catalytic activity and new functions. Different prebiotic models (Eigen, 1971; Prigogine et al., 1972; Babloyantz, 1972; and Goldbeter, 1973), have made it possible to show that:

1. A thermodynamic threshold exists for relatively simple polymerization mechanisms, leading from one stationary state, characterized by low polymer density, to another stationary state with high polymer density.

ISBN 0-201-03438-7/0-201-03439-5pbk.

2. Above this threshold, there is the wealth of possible behavior characteristic of dissipative structures (limit cycle, chemical waves, etc.).
3. When the appearance of new chemical substances corresponds to a new function, the specific dissipation (e.g., per unit mass) of the system increases at the point of instability. The new regime corresponds to a higher level of interaction between system and environment. This behavior has been called *evolutionary feedback* (Babloyantz, 1972). Indeed, in increasing the dissipation, the class of fluctuations leading to instabilities is widened.

These considerations have also been applied to ecological evolution. (See the Appendix to this chapter and Allen 1975, 1976.)

5. BIOLOGICAL APPLICATIONS: AGGREGATION

The concepts which we have developed may now be illustrated by some biological examples chosen for their close relation to social phenomena. Let us begin with the phenomenon of aggregation in slime molds, a species of *Amoeba*. The life cycle of these microorganisms is represented in Figure 5.7. When the food has been used up, the amoeba start aggregating by moving toward attractive centers that seem to form spontaneously. A mobile mass containing from 10 to 10^5 cells—the pseudoplasmodium—forms. This mass changes shape and becomes a body composed of two structures: a foot, or base, whose cells are rich in cellulose; and a round mass above, rich in polysaccharides. This is a true differentiation or structuration. This phenomenon develops during a period of between 20 and 50 hours.

Here, we shall study only the aggregation process, whose evolution is represented in Figure 5.8. It has been shown that it is the attraction exerted by a chemical substance, acrasin, on the amoebas which is at the origin of the aggregation: this type of behavior is called *chemotaxis*. It has been possible to establish that the chemical identity of acrasin is a cyclic compound C-AMP, a substance that plays an important role in many biochemical processes. During the aggregation, there are no cell divisions, and the total number of amoebas is conserved. The density variations of the amoebas may be expressed in terms of two types of displacements: one term for random motion represented by Fick's law, and another for the chemotactic displacement, related to the gradient of acrasin. Therefore, the evolution equation for $a(r, t)$ may be written

$$\frac{\partial a}{\partial t} = - \nabla \cdot (D_1 \nabla_\rho) + \nabla \cdot (D_2 \nabla_a) \qquad (5.1)$$

where ρ is the density of acrasin, D_1 the chemotactic coefficient, and

ISBN 0-201-03438-7/0-201-03439-5pbk.

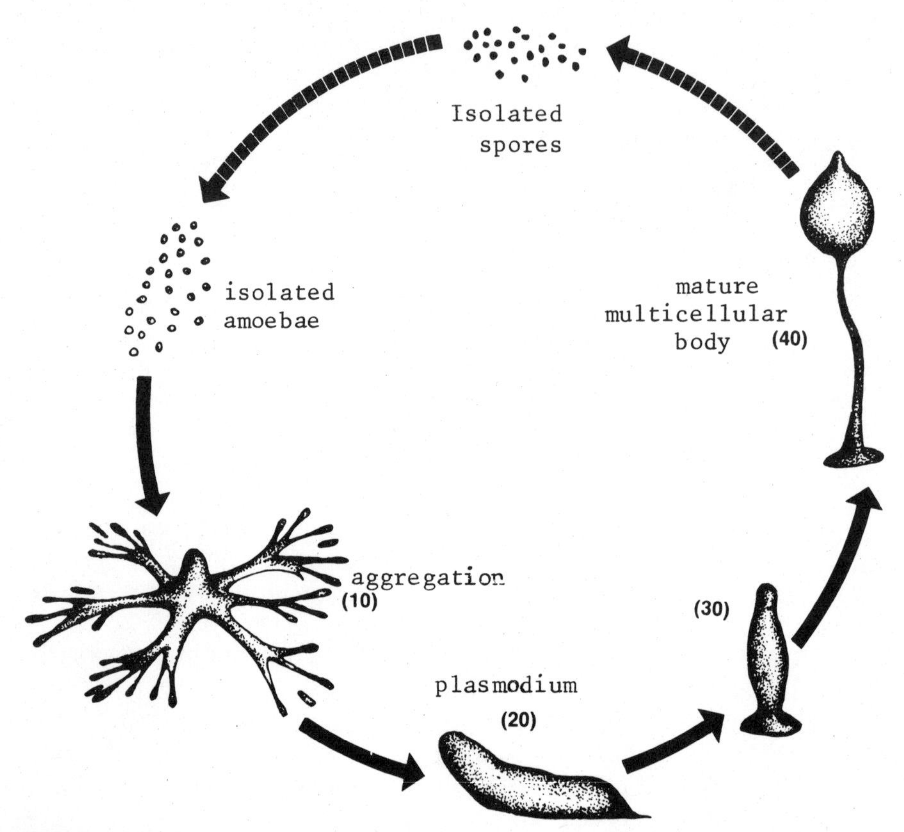

FIGURE 5.7. The life cycle of the slime mold. Numbers in parentheses indicate the hours elapsed from the moment that cell division ceases in the free phase (see Susman, 1964).

D_2 the diffusion coefficient. It is generally supposed that D_1 is a term of the form

$$\frac{\delta \cdot a(r, t)}{\rho} \tag{5.2}$$

where δ is some constant. In addition to chemotaxis we have an enzymatic reaction. The slime molds release into the medium an enzyme, acrasinase (η), which destroys acrasin. The corresponding reaction is

$$\rho + \eta \underset{k_{-1}}{\overset{k_1}{\rightleftharpoons}} C \overset{k_2}{\rightleftharpoons} \eta + \epsilon \tag{5.3}$$

ISBN 0-201-03438-7/0-201-03439-5pbk.

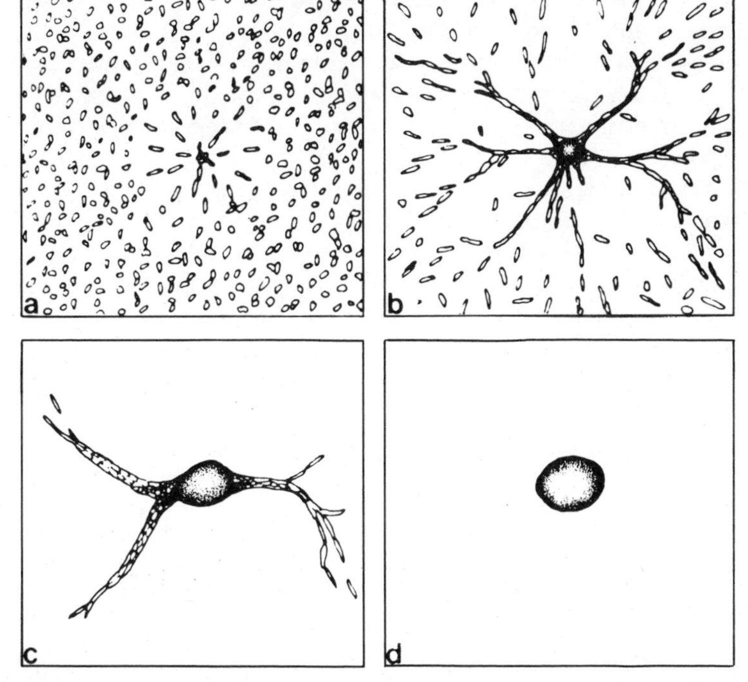

FIGURE 5.8. The evolution of the aggregation process in the life cycle of the slime mold (see Susman, 1964).

where C is an intermediate complex and ϵ the product of the degradation. If we suppose that the total concentration of the enzyme (free and in the complex) is a constant, the problem of aggregation reduces to two pairs of equations (Keller and Segel, 1970):

$$\frac{\partial a}{\partial t} = - \nabla \cdot (D_1 \nabla_\rho) + \nabla \cdot (D_2 \nabla_a) \tag{5.4}$$

$$\frac{\partial \rho}{\partial t} = - k(\rho)\rho + af(\rho) + D_\rho \nabla_\rho^2 \tag{5.5}$$

where D_ρ is the diffusion coefficient of the acrasin; $k(\rho) = (\eta_0 k_2 k)/(1+k_\rho)$; $k = k_1/(k_{-1} + k_2)$; and $af(p)$ is the amount of acrasin produced by the amoebas.

These equations always admit a stationary homogeneous solution corresponding to the thermodynamic branch. One can determine the conditions which would render this state unstable. Instability is favored if the cells undergo a sufficient augmentation of their sensitivity to a given gradient of acrasin, that is, if δ increases sufficiently. An increase

ISBN 0-201-03438-7/0-201-03439-5pbk.

in the rate of production of the acrasin [the term $\nabla^2 \rho$ of (5.5)] or of its concentration in the medium, contributes to the destabilization of the homogeneous stationary state. A certain critical wavelength exists which essentially determines the spatial distribution of the aggregates. The principal predictions of the model have been found to be in agreement with experiment (Keller and Segel, 1970).[2]

We have here an excellent example of order by fluctuation. The fluctuations in the amoeba distribution destroy the uniform configuration and lead the system finally to a new *inhomogeneous* distribution. The analogy with the formation of towns starting from a uniform population is obvious.

6. SOCIAL INSECTS

The relations between individuals of a species vary considerably within the animal kingdom. Certain groups limit themselves to sexual behavior or to the struggle for the defense of their territory. Among insects, social organization attains a maximum complexity with the Hymenoptera and the termites: the survival of an individual is practically impossible outside the group (Wilson, 1971). The interactions between individuals are physical: sound, vision, touch, and the transmission of chemical signals. The regulation of the castes, nest construction, formation of paths, and the transport of material or of prey are different aspects of the order reigning in the colony. In the following, we discuss two striking examples from our point of view (see also Deneubourg and Nicolis, 1976).

Collective Movements of Ants

Ants, like insects in general, synthesize a great many chemical substances (pheromones), which regulate their behavior. The "path" pheromones mark on the ground the direction of food sources or of the nest. The substance laid down can diffuse into the neighboring space. A "tunnel" of pheromone is thus created, centered on the axis of displacement of the insect who laid it. His fellows have a tendency to follow the same direction at the place where the density of pheromone molecules reaches a maximum. With certain groups, such as soldier ants, one may observe the collective movement of several thousand individuals. Macroscopic

[2] Propagating waves have been observed during aggregation by Gerisch and Hess (1974). Their mechanism has been discussed recently by Goldbeter (1975).

ISBN 0-201-03438-7/0-201-03439-5pbk.

structures appear and vary in form from species to species (Rettenmeyer, 1963, Hangartner, 1967). (See Figure 5.9.)

We formulate now the equations of change. Let $C(r,\theta)$ and $H(r,\theta)$ be the insect and pheromone concentrations, respectively, expressed in polar coordinates r,θ. Suppose that the nest is at $r = 0$. The following hypotheses are used in writing the kinetic equation of the pheromone:

1. The ants emit a quantity α of pheromone per unit of time.
2. H decomposes at a rate proportional to its density: $-\beta H$.
3. Its propagation in the medium obeys Fick's law where D_H is the diffusion coefficient.

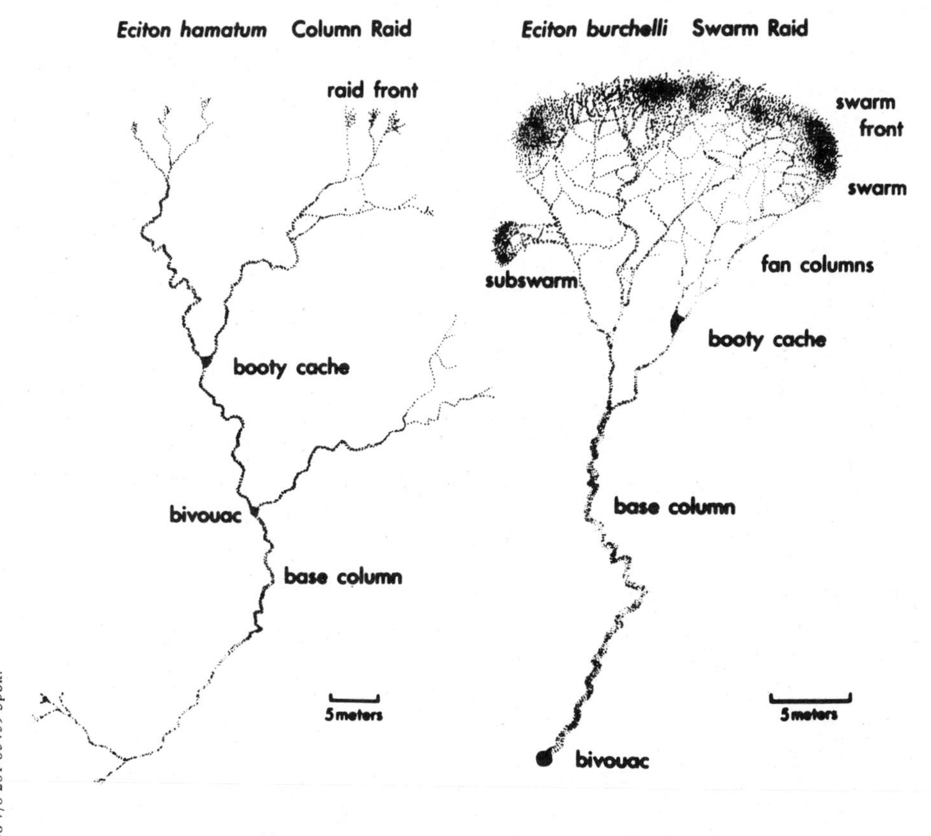

FIGURE 5.9. Collective movement of ants: Two types of structure, referring to different species (see Rettenmeyer, 1963).

ISBN 0-201-03438-7/0-201-03439-5pbk.

In this way we obtain

$$\frac{\partial H}{\partial t} = \alpha C - \beta H + D_H \left[\frac{1}{r} \frac{\partial}{\partial r} \left(r \frac{\partial H}{\partial r} \right) + \frac{1}{r^2} \frac{\partial^2 H}{\partial \theta^2} \right] \tag{6.1a}$$

Similarly, the kinetic equation for C is

$$\frac{\partial C}{\partial t} = H_r C + \frac{D_\theta}{r^2} \frac{\partial^2 C}{\partial \theta^2} + \frac{\gamma}{r^2} \frac{\partial}{\partial \theta} \left(C \frac{\partial H}{\partial \theta} \right) \tag{6.1b}$$

The first term on the right-hand side in Eq. (6.1b) contains the radial component of the collective movement. As described earlier, the orientation in the pheromone gradient occurs perpendicularly to the principal direction. The angular dependence has two components: one corresponding to diffusion according to Fick's law (second term on the right-hand side) and the other to the attraction of the pheromone. D_θ is the diffusion coefficient of C, which depends on θ, and γ is the chemotactic coefficient.

There exists a critical value γ_c for which a stationary homogeneous solution becomes unstable. The system then evolves to an inhomogeneous stationary state depending on θ, with a critical wavelength linked to the different parameters of the model. Accordingly, different branching structures will appear according to these parameters, as observed in different ant societies.

Construction of a Termite Nest

Social insects are also characterized by coordinated behavior such as nest construction. The scale of such constructions far exceeds the size of an individual. It is interesting that for certain species the population of a colony attains several million individuals and the total weight of a termite nest can sometimes attain several tons. In spite of this size, such a structure can result from a simple behavioral pattern of the insects. Gallais-Hamonno and Chauvin (1972) have simulated the construction of the dome of an ant nest.

In the first stage of the construction of a termite nest the insects raise a group of pillars and walls, which, if sufficiently close, are joined to form arches. Afterwards, the space between the pillars is blocked. Grassé (1959), in particular, has studied this first stage, and his observations have led him to formulate the theory of "stigmergy," which expresses the "interaction" between insects and work. Summarizing

ISBN 0-201-03438-7/0-201-03439-5pbk.

the observations of Grassé, we may conclude that there exist two phases: (1) an uncoordinated phase; and (2) a coordinated phase. The uncoordinated phase is characterized by the random deposition of building material. Many small deposits are thus distributed on the surface available to the insects. When one of these deposits becomes sufficiently large, the second phase starts. On the aggregation of matter the termites now deposit even more material, but in a preferential way. A pillar or a wall grows according to the initial disposition of the deposit. If these units are isolated, construction stops; but if they are close to each other, an arch will result (see Figure 5.10).

Different types of stimulus, principally chemical but also mechanical, intervene. Let us consider how these facts can be expressed by using a simple model: The existence of a deposit at a specific point stimulates the insects to accumulate there more building material. This is an autocatalytic reaction. A model containing this autocatalytic factor, together with the random displacement of insects, is formally similar to the examples in Section 3.

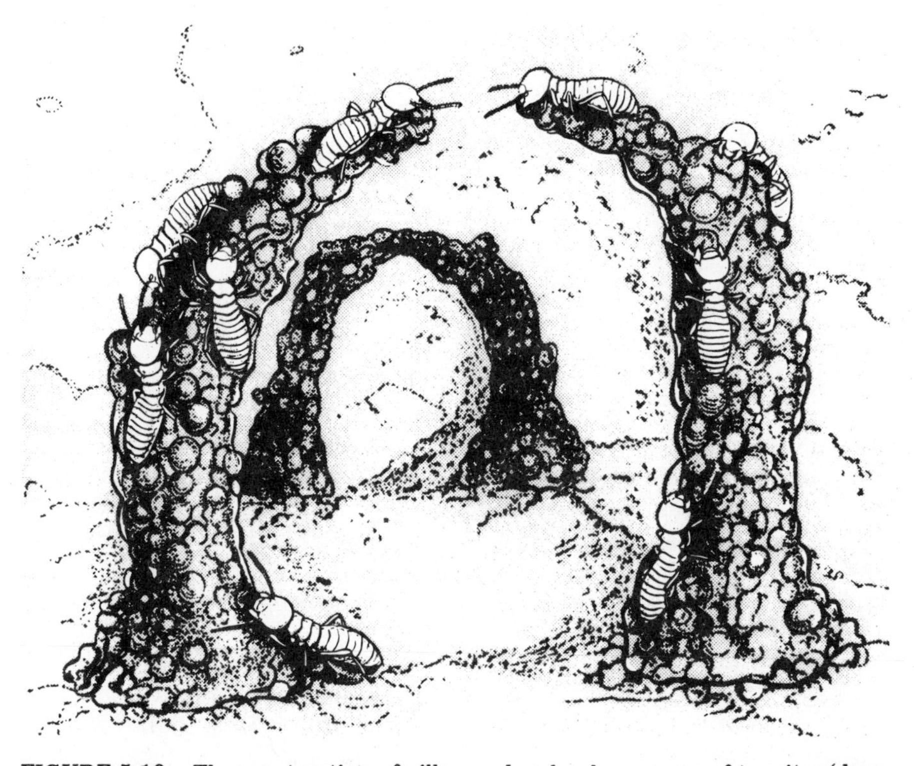

FIGURE 5.10. The construction of pillars and arches by a group of termites (drawing by Turid Holldobler, see Wilson, 1971).

ISBN 0-201-03438-7/0-201-03439-5pbk.

The termites, in manipulating the construction material, give to it a particular scent, which diffuses in the atmosphere and attracts the insects toward the points of highest density, where deposits have already been made. Let C be the concentration of insects carrying material (P). The deposition of the material is favored by the quantity of P present, by means of H (scent), and is supposed to be proportional to C. In addition, we suppose that P loses its attractive capacity at a speed proportional to its density. We then have for P

$$\frac{\partial P}{\partial t} = k_1 C - k_2 P \tag{6.2}$$

The equation for the evolution of the scent H contains a production term proportional to the density P. Its decomposition rate is of the form $-k_4 H$. Fick's law represents its spatial propagation.

$$\frac{\partial H}{\partial t} = k_3 P - k_4 H + D_H \frac{\partial^2 H}{\partial r^2} \tag{6.3}$$

where D_H is the diffusion coefficient. We also have to consider a similar equation for C. This includes a flow Φ of insects carrying material $-k_1 C$. In addition, we include diffusion and motion directed toward the sources of the scent.

This leads to the equation

$$\frac{\partial C}{\partial t} = \Phi - k_1 C + D \frac{\partial^2 C}{\partial r^2} + \gamma \frac{\partial}{\partial r} \left(C \frac{\partial H}{\partial r} \right) \tag{6.4}$$

where γ is the chemotactic coefficient.

The concept of the instability of the homogeneity and the role of the fluctuations are here again well illustrated. The uncoordinated phase, to use Grassé's terminology, corresponds to the homogeneous solution of Eq. (6.1), (6.3), and (6.4). From a slightly larger deposit somewhere, or in other words from a sufficiently large fluctuation, a pillar or a wall can appear. This stage corresponds to the amplification of the fluctuation. Order therefore appears through fluctuations. The regularity of the structure spreads and is characterized by a wavelength which is a function of the different parameters of the model.

This model leads to a certain number of conclusions of experimental interest. It becomes possible to perform numerical experiments reproducing Grassé's observations for diverse conditions, such as different insect or scent densities. The role of the "fluctuations," which are given here by the initial deposits, can also be studied by introducing decoys into the system (Grassé, 1959).

ISBN 0-201-03438-7/0-201-03439-5pbk.

The mathematical aspects of works of art, particularly architecture, have been emphasized by many authors. It is remarkable to find this aspect already in constructions built by social insects.

7. THE FORMATION OF DISSIPATIVE STRUCTURES—THE STOCHASTIC DESCRIPTION

So far, we have discussed dissipative structures according to a deterministic description. This method is valid as long as the fluctuations, which is to say the deviations from average values, remain small. Equilibrium statistical mechanics tells us that the order of magnitude of the fluctuations in a system with N degrees of freedom is $N^{1/2}$ (see Glansdorff and Prigogine, 1971; and McQuarrie, 1967). The relative importance of the fluctuations therefore tends to become zero for $N \to \infty$.

Near an instability, however, the importance of the fluctuations can become crucial. In order to take into account these fluctuations, the simplest procedure is to suppose that the chemical reactions lead to a Markov process of "birth and death" in the space defined by the particle numbers X_i of the different chemical species i. This makes it possible to establish a "master" equation which gives the evolution of the probability P of finding given values of the particle numbers at time t. For example, let us consider the two monomolecular reactions

$$A \xrightarrow{k_1} X \xrightarrow{k_2} F$$

Let $P(A, X, F, t)$ be the probability of the values A, X, F of the particle numbers at time t. In order to establish a master equation, one must consider all the transitions that could lead, during the interval $t \to t + \Delta t$, from one state A, X, F to another, and vice versa. Take the first reaction,

$$A \xrightarrow{k_1} X.$$

In order to have the state (A, X, F) at time $t + \Delta t$, it is necessary that the state have been at time t $(A + 1, X - 1, F)$; in consequence

$$\begin{aligned} P(A,X,F,t+\Delta t) = {} & (\text{probability of reaction}) \times P(A+1,X-1,F,t) \\ & + (\text{probability of nonreaction} \times P(A,X,F,t) \end{aligned} \tag{7.1}$$

ISBN 0-201-03438-7/0-201-03439-5pbk.

The probability of a reaction toward the state (A, X, F) is given by

$$\text{probability of reaction} = k_1 A\ \Delta t \tag{7.2}$$

from which follows

$$\begin{aligned} P(A,X,F,t+\Delta t) &= k_1(A+1)\ \Delta t\ P(A+1,X-1,F,t) \\ &\quad + (1-k_1 A\ \Delta t)P(A,X,F,t) \end{aligned} \tag{7.3}$$

The equation of evolution of $P(A, X, F, t)$ results by adding the contributions of both reactions and taking the limit $\Delta t \to 0$:

$$\begin{aligned} \frac{dP}{dt} &= k_1(A+1)P(A+1,X-1,F,t) - k_1 AP(A,X,F,t) \\ &\quad + k_2(X+1)P(A,X+1,F-1,t) - k_2 XP(A,X,F,t) \end{aligned} \tag{7.4}$$

or in a more general form,

$$\frac{dP}{dt}\left(\{X_i\},t\right) = \sum_{\{X_i'\}} W\left(\{X_i'\},\{X_i\}\right) P\left(\{X_i'\},t\right) \tag{7.5}$$

P is the probability function and W the probability of transition, per unit time, between states $\{X_i'\}$ and $\{X_i\}$.

Solving Eq. (7.5) for small fluctuations around the equilibrium state shows that such fluctuations obey a Poisson distribution. This implies that the mean square deviation $<\delta X^2>$ is equal to the mean value:

$$<\delta X^2> = <X> \tag{7.6}$$

However, Eq. (7.5) also allows us to study the behavior of fluctuations in systems maintained outside equilibrium, by imposing fixed values on the concentrations of the initial and final products. It is the necessary that there be a clear separation of the time scales of the "fluctuating system" and the "outside world" (Nicolis and Babloyantz, 1969). We note, however, that this approach considers the fluctuations to be spread homogeneously throughout the whole reacting volume. The results of this work have been surprising. We have shown that it is only in the case of linear systems that the small fluctuations can be described by Poisson's formula (see Nicolis and Prigogine, 1971; Nicolis et al., 1974; and Prigogine et al., 1975). For nonlinear systems far from equilibrium, the distribution law of the fluctuations of the reacting substances is not that of Poisson. How can this be explained? Recent studies (Mazo, 1975)

ISBN 0-201-03438-7/0-201-03439-5pbk.

have shown that there is an unexpected aspect that has to be considered. The fluctuations are local events, and one must consider a supplementary parameter scaling the extension of the fluctuations. This will be a new characteristic length determined by the intrinsic dynamics of the system and independent of the dimensions of the reacting volume. Thus, there is an essential difference in the behavior of the fluctuations depending on their spatial extension. Only fluctuations of sufficiently small dimensions obey Poisson statistics. This is a very important result because it implies that, conversely, only fluctuations of a sufficient extension can attain enough importance to compromise the stability of the macroscopic state considered.

Thus, our recently developed theory leads quite naturally to the notion of a *critical fluctuation* as a prerequisite for the appearance of an instability.

These conclusions can be illustrated with the help of a simple model (see Figure 5.11). Consider a volume V inside which chemical reactions are taking place. Let ΔV be a small subvolume of this system.

Besides the chemical reactions, this small subvolume is linked to the big volume V, the environment, by exchanges of matter. Let $\mathscr{D}$ be the corresponding coefficient of transport, and X_{in} and X_{ex} the characteristic composition in ΔV and in the region $\Delta\epsilon$ at the interface of V and ΔV. A master equation for the total system may be written

$$\begin{aligned}\frac{d}{dt}P(X_{in},X_{ex},t) &= R(X_{in} + R(X_{ex}) \\ &+ \mathscr{D}\,[(X_{in}+1)P(X_{in}-1,X_{ex}+1,t) - X_{ex}P(X_{in},X_{ex},t)] \\ &+ \mathscr{D}\,[(X_{in}+1)P(X_{in}+1,X_{ex}-1,t) - X_{in}P(X_{in},X_{ex},t)]\end{aligned} \tag{7.7}$$

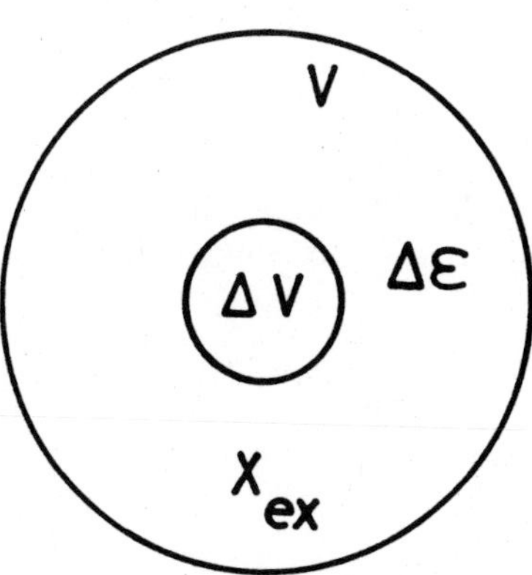

FIGURE 5.11. Coupling by the exchange of matter between the small subvolume ΔV and its environment.

ISBN 0-201-03438-7/0-201-03439-5pbk.

where $R(X_{in})$ and $R(X_{ex})$ represent the contributions of the chemical reactions, while the other terms are linked to the transfer of matter between ΔV and its environment.

By considering the following hypotheses:

1. the chemical reaction responsible for the instability occurs in the same way in V as in ΔV;
2. the interaction with the surrounding system occurs according to the average state of the outside medium;
3. the transfers between ΔV and the exterior depend on the instantaneous state of ΔV;

we are led to an equation concerning only the probabilities inside the subvolume, $P(X, t)$, $(X \equiv X_{in})$:

$$\frac{dP(X,t)}{dt} = R(X) + \mathscr{D} \langle X \rangle [P(X-1,t) - P(X,t)] + \mathscr{D} [(X+1)P(X+1,t) - XP(X,t)] \tag{7.8}$$

with

$$\langle X \rangle = \sum_{X=0}^{\infty} X\,P(X,t) \tag{7.9}$$

This *nonlinear master equation* expresses the competition between the chemical reactions, which tend to augment fluctuations, and the transfers of matter, which tend to damp them by homogenization. It is only when these two types of terms are of the same order of magnitude that an instability can manifest itself.

Let us now examine more closely what the implications of the new approach are for a system which crosses the threshold of instability. At the critical point separating the stable region from a regime which amplifies the fluctuations, the transfer parameter $\mathscr{D}$ will be linked to the chemical parameters on the one hand and to the dimension of the subvolume ΔV on the other. A qualitative estimate of this gives[3]

$$\mathscr{D} = \frac{D}{\ell^2} \tag{7.10}$$

where D is Fick's diffusion coefficient and ℓ a characteristic length of ΔV.

[3]For the particular case of dilute gases Nicolis et al. (1974) have introduced $\mathscr{D} = D/\ell\ell_c$ where ℓ_c is the mean free path in the gas.

ISBN 0-201-03438-7/0-201-03439-5pbk.

By introducing (7.10) into the condition for instability, a relation can be established between the size of the subsystem considered and the rate of amplification of the fluctuations.

This calculation has been performed for a particular chemical example: the establishment of a limit cycle in the trimolecular mechanism (Nicolis and Prigogine, 1971). The results are shown in Figure 5.12, which clearly shows the competitive action of the chemical parameters and the dimension of the perturbation on the stability of the system. Three domains, having different stability properties, appear:

1. A stable region where all fluctuations will be damped, whatever their dimension.
2. A region where $k > k_c$ but where the length is shorter than the critical length; the fluctuations here are also damped.
3. A region where both $k > k_c$ and $\ell > \ell_c$; the fluctuations are amplified and invade the whole system.

The result of this work is to show that dissipative structures form by a *nucleation process.* Fluctuations on a sufficiently small scale are always damped by the medium. Conversely, once a fluctuation attains a size beyond a critical dimension, it triggers an instability.

Although this theory is very recent, there are several arguments which point to its validity. First, certain calculations performed by a computer

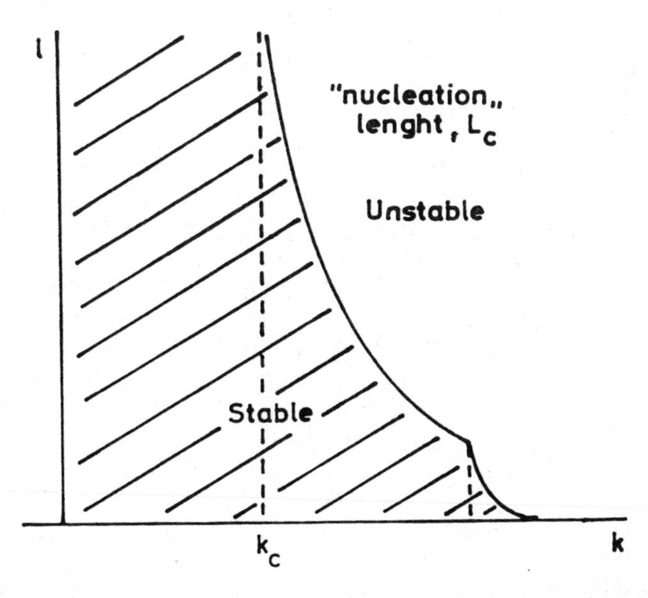

FIGURE 5.12. The coherence length l of a fluctuation at the critical point, as a function of the characteristic chemical parameter k; k_c is the critical value of k above which the system is unstable from the macroscopic point of view.

ISBN 0-201-03438-7/0-201-03439-5pbk.

have shown the role played by the dimension of the fluctuations (Mazo, 1975; Portnow, 1974). Furthermore, experiments confirm, at least qualitatively, the existence of a critical size of the fluctuations in the triggering of chemical or hydrodynamic instabilities (Nitzon et al., 1974; Pacault et al., 1975).

From the point of view of basic principles, the properties of this new class of stochastic equations are very remarkable, because they show the restraining role of the environment.[4] Clearly, it is an essential factor in the possible sociological applications and it is precisely to these questions that we shall turn in the final section of this chapter.

8. DISSIPATIVE STRUCTURES AND SOCIAL SYSTEM

The theory of dissipative structures lends itself to a description of the self-organization of matter in conditions far from thermodynamic equilibrium. These structures have a coherent character linking their mechanism ("chemical reactions") to their spatiotemporal organization. Furthermore, we have seen that such systems present both a deterministic character, described by kinetic equations of type (3.2) or (6.2)-(6.4) and stochastic character (fluctuations) described by nonlinear stochastic equations such as (7.5) This theory also links the three levels of description

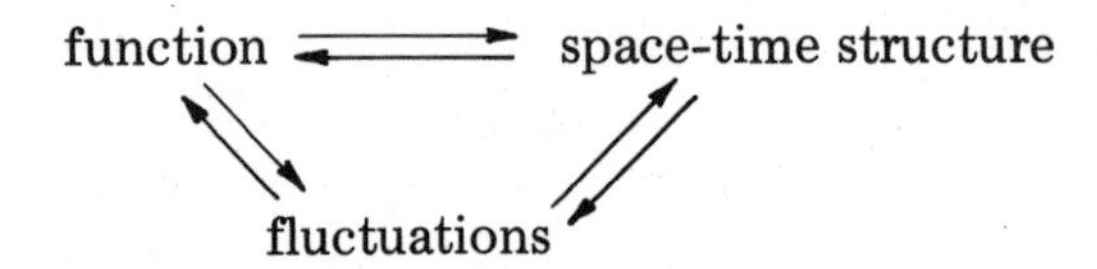

The theory of dissipative structures has been applied with great success to biological problems (see Sections 5 and 6). It must be recalled that even in the simplest cells the normal metabolic processes imply several thousand complex chemical reactions. Therefore, out of absolute necessity all these processes must be coordinated. These coordinating mechanisms constitute an extremely sophisticated functional order. Thus, biological order is both functional and spatiotemporal order.

It is therefore tempting to apply these concepts to problems of social structure. There, as in the case of biological structures, the functional

[4] This approach also spotlights the weaknesses of the traditional methods (Poisson processes, Markov chains) when applied to nonlinear processes far from equilibrium (Malek-Mansour and Nicolis, 1975). These classical methods do not lend themselves to modeling of the characteristic phenomena occurring within a society (see Section 8).

ISBN 0-201-03438-7/0-201-03439-5pbk.

aspect is associated with specific structures. From one particular point of view we are in an even more favorable situation than in biology, because life is a very ancient phenomenon and its origin must imply a very considerable sequence of successive instabilities. In contrast, we are at least partially informed on the development of societies, thanks to archeological and ethnological evidence (see, e.g., Leroi-Gourhan, as quoted by Janne, 1963; and Rachet, 1969). In particular, we have considerable information concerning the history of technology and of the tools of primitive societies and about the way in which material progress has been reflected in the organization of society.

Mathematical models have often been introduced in the study of social and ecological phenomena. We may cite here, for example, the model of ecological competition of Volterra and Lotka (see Goël et al., 1971) or the kinetic theory of traffic flow (Prigogine and Herman, 1971). It is also well known that in spite of its limitations, the theory of Markov chains has found numerous applications in the social domain (Weidlich, 1974).

The important point is that we can now go further. The perfecting of the mathematical tools [theory of bifurcations, structural instabilities (see Section 4), and nonlinear stochastic equations] permits us to discuss in a more precise way some basic concepts introduced by sociologists.

As an example, we shall consider the notion of the "quantum of action" to which Henri Janne (1963), in his monograph on sociology, attaches great importance. He writes (p. 42):

> It is convenient here to introduce the notion of a "quantum of action." The quantum of action of a factor must be sufficient for the factor to be taken into consideration. Below a certain threshold, the factor has no effect (insufficient quantum of action). It can be "dominant" when its action renders all other factors negligible . . . . The dominant quantum of action joins the game of classical causality.

The concept introduced by Janne is analogous to the critical fluctuation which we have discussed in Section 7. With fluctuations below the critical threshold, the system returns to its initial state. Beyond the threshold, it evolves to a new structure. The appearance of a critical fluctuation leads in this sense to a deterministic evolution (whence Henri Janne's reference to classical causality).

Althusser (Althusser and Balibar, 1973) has expressed very well the necessity of clarifying the "epistemological" significance of the new concepts introduced by the founders of modern sociology. Speaking of theoretical problems introduced by Marxist ideas, he says (p. 61),

ISBN 0-201-03438-7/0-201-03439-5pbk.

> By what concept can one think of the new type of determination, only recently identified, where for example the phenomena occurring in a region are determined by the structure of that region? More generally, by means of what concept, or group of concepts, can one conceive the determination of the elements of a structure and the structural relations between them, as well as all the effects of these relations, in terms of the efficiency of this structure? And, *a fortiori*, by what concept, or concepts, can one imagine the determination of a subordinate structure by a dominant one? In other words how can one define the concept of a structural causality?

The origin of modern sociology has often been attributed to those thinkers of the nineteenth century who defined and forged the concepts on which theoretical sociology is based. In contrast, other sociologists attribute it to the founders of social statistics, such as Quetelet (see Aron, 1967).

The development of appropriate statistical mathematics capable of handling the complexity of social problems could certainly serve as a bridge between these two complementary ways of envisaging sociology. It is in this perspective that I should like to present some remarks:

(a) Social phenomena are described by nonlinear equations. This results directly from their relational or social character. This relational character appears under different titles in the works of the founders of sociological thought. Tarde (1890) speaks of imitation, Durkheim (1973) of solidarity. The mathematical transcription of this element leads precisely to this nonlinear aspect. We have already cited the nonlinear Volterra-Lotka equations used currently in ecology. Similarly, in the dynamics of phenomena of consumer choice appear different contributions, some "linear," corresponding to "individual" decisions, and others "nonlinear," corresponding to decisions taken under the influence of the environment (friends, mass media, etc.). Finally, a well-understood example is that of vehicular traffic flow, where the interaction between drivers leads precisely to the nonlinearity in the integrodifferential equations for the distribution function of the velocities of the drivers.

(b) The coherent behavior of a society has been underlined many times; in fact, the relation between structure and function is so apparent that it seems unnecessary to stress it further.

(c) Several research workers, notably Gregory Bateson (cited by Janne, (1963, p. 117), have felt the importance of emphasizing change in the description of social systems as coherent systems characterized by the structure-function relation. Bateson introduces the notion of dynamic equilibrium, with the help of which he shows that "any social system, in spite of its static appearances, contains at least small amplitude changes, appearing continuously and compensating one another." This observation is to be compared to the existence of dysfunctions underlined by Janne (1963, p. 111). In our description, these phenomena correspond to the existence of fluctuations inherent

ISBN 0-201-03438-7/0-201-03439-5pbk.

to the statistical description. The deterministic description refers only to averages.

(d) The social fact finds its expression in constraints imposed upon an individual. This constraint appears in our theory as the nonlinear terms in the stochastic equation (7.7). It tends to stabilize the system with respect to certain fluctuations, such as the dysfunctions mentioned in (c) above.

In summary, our description includes two complementary aspects. These aspects correspond essentially to the dialectic between mass and minority, to use the expression introduced by F. Perroux (1964). The first aspect is that of an average behavior (the "average" man of Quetelet), the second is the existence of fluctuations which, when they exceed a critical level, influence that average because they drive the system to a new average state.

We believe that our model contains certain of the indispensable elements for the building of a theoretical sociology which cannot neglect either of these two aspects without seriously altering the significance of the social system.

In the following, we shall make some remarks which illustrate the importance of the preceding considerations. We do not seek here to formulate a precise mathematical model of a given social activity. Any such model would necessarily imply a detailed discussion of the parameters in play and would fall outside the range of this chapter. We include, however, in the Appendix a summary of some recent work by P. M. Allen.

A first point is the problem of the very existence of societies. Is there a limit to complexity? The question has been discussed many times in the literature. An excellent exposition is given in the monograph by May (1973). The more elements there are in interaction, the higher the degree of the secular equation determining the characteristic frequencies of the system (see Section 4). The greater will therefore also be the chances of this equation's having at least one positive root and hence indicating instability.

Several authors have suggested that historical evolution selects certain particular types of systems that are stable. It has nevertheless proven difficult to give a quantitative form to such a suggestion. Our approach leads to a different answer: A sufficiently complex system will generally be in a metastable state (see also Holling's discussion of the same point in Chapter 4, Section 3). The value of the threshold of metastability depends on the size of the coefficient appearing in the stochastic equation (1.7). This coefficient, as we have seen, is a measure of the coupling of the fluctuating system with the outside world. This point of view seems to be in agreement with the one held by sociologists who conclude that a society has a limited power of integration. If the per-

ISBN 0-201-03438-7/0-201-03439-5pbk.

turbation exceeds that power of integration, the social system is destroyed or gives way to a new organization.

The existence of the constraint is indispensable in order to distinguish between an "average state" (including periods of development) and fluctuations leading to a new state. One may think that in a complex society the possibilities for instability (resulting, e.g., from new inventions) always exist. However, only certain inventions will go beyond the individual domain to a domain of integration with society. One has only to think of the wheel, used in the pre-Columbian epoque as a toy, but not as a means of transport.

The formalism that we have obtained leads quite naturally to a preliminary classification of societies according to the following two parameters:

(1) "Complexity," measured by the number of interacting functional elements;
(2) social pressure, measured by the parameter $\mathcal{D}$.

It is of particular interest to consider the two limiting cases (b) and (c) in Table 5.1. Case (b) corresponds to a simple, "conformist" society. One thinks here of the archaic social systems which Lévi-Strauss has compared with "clocks" (Charbonnier, 1969). The opposite case is that of a complex system with a feeble coherence corresponding to historical societies which Lévi-Strauss compares to "steam engines." One should notice that the nonhistoric nature, or "crystallinity," of certain archaic societies corresponds, from this point of view, to active repression of the fluctuations (see also Chapter 8 by Taylor).

TABLE 5.1

Complexity	*Social Pressure*	*Stability*
-	-	(a) ?
-	+	(b) Stable
+	-	(c) Unstable
+	+	(d) ?

All other things being equal, the repression is stronger, the smaller the fluctuating group in which it acts (see Section 7). This is probably related to the remark of Gurvitch (quoted by Janne, 1963, p. 344): the family (a relatively small unit) is a conformist element in a society containing small fluctuations. In contrast, society as a whole constitutes the greatest dimension of a fluctuation, and in consequence is subject to the least constraint (except at certain times, such as during a war). Accordingly, ethnologists have been able to identify a great number of distinct societies (see also Chapter 10 by Maruyama). Does this mean that the evolution

ISBN 0-201-03438-7/0-201-03439-5pbk.

of societies is not subject to any general rule? In 1922, Lotka formulated his law of maximum energy flow (Lotka, 1956). In thermodynamic terms, this corresponds to a law of increase of entropy production per individual. This law seems to agree with the laws of technological evolution. As Leroi-Gourhan (quoted by Janne, 1963, p. 288) has written: "In the technical domain, the only features which will be transmitted are those which represent an improvement in the procedures. One may adopt a language which is less supple, a religion which is less developed, but one will never exchange a plough for a hoe." The plough leads necessarily to an augmentation of the exploitation of natural resources and in consequence a greater energy consumption per individual. It is interesting to compare this tendency with the entropy production which appears in the early stages of embryonic life (Zotin, 1972). The increase of entropy production in turn renders possible the appearance of new instabilities. We have already pointed to the evolutionary feedback in Section 4 of this chapter.

This increase in entropy production is related to the effect of structural instabilities discussed in Section 4. There is a close analogy between the "invention" of new techniques and the structural instability leading to new chemical mechanisms. Of course, there is no question of classifying societies according to a single criterion such as their energy or entropy production. This is merely a characteristic of evolution, but a very important one because of its universality. In contrast, our approach clearly shows the rather oversimplified nature of theories of "progress" (linear progress, cycles, etc.).

The ideas of "infrastructure" and "superstructure" have given rise to interminable discussions (see, e.g., Aron, 1967). It seems worthwhile, therefore, to indicate that within the framework of our formalism, these ideas take on a very direct meaning. A structural instability may result from the occurrence of a new function arising from a fluctuation. With such a fluctuation, one may associate a modification of the infrastructure. The relation between the space-time function-structure will be modified if the fluctuation leads the system to a new dissipative structure. From this point of view, the space-time structure appears as the "superstructure."

Of course, the very possibility of fluctuations depends on the restraining character of a society, and therefore of the "superstructure." The notions of average stage and of fluctuations can only be defined with respect to one another.

As we have already indicated, it is necessary to separate the development periods from the periods of instability which lead to new structures. The problem of forecasting is entirely different in the two cases. In the former, it suffices essentially to study deterministic laws, whereas

ISBN 0-201-03438-7/0-201-03439-5pbk.

in the latter this is certainly not so. It has often been said that the life of the average man in Europe in the eighteenth century was very similar to that of the average man in the developing countries today. In the eighteenth century, however, fluctuations, triggered by the development of the sciences, were already growing. It is in the nineteenth century that we see these fluctuations attaining the "average" state and constituting a force which modified the destiny of European societies in their entirety. It is not surprising, therefore, that it was at this moment that the problem of time, of history, became the central theme of epistemology. Auguste Comte summed this up by predicting, "Our present century will be principally characterized by the irrevocable preponderance of history in philosophy, in politics and even in poetry."

9. CONCLUSIONS

Bergson (1963, p. 503) made the following statement: "The further we penetrate the analysis of the nature of time, the more we understand that duration signifies invention, creation of forms and the continual elaboration of what is absolutely new."

We recognize that we are beginning to clarify these notions of "invention" and "elaboration of what is absolutely new" by the mechanism of successive instabilities caused by critical fluctuations (Prigogine, 1973). The discovery of such mechanisms, which play such an essential role in a vast domain stretching from physics to sociology, is obviously a preliminary step toward some harmonization of the points of view developed in these different sciences.

ACKNOWLEDGMENTS

This chapter was written with the active participation of our groups at Brussels and Austin, and especially of Professor Nicolis and Drs. Lefever, Allen, and Deneubourg. It was finalized during a stay at the General Motors Research Center, and I wish to thank Dr. R. Herman and A. Butterworth for stimulating discussions, and Dr. P. Chenea for his interest.

ISBN 0-201-03438-7/0-201-03439-5pbk.

Appendix

POPULATION DYNAMICS AND EVOLUTION

Peter Allen

The principle of order through fluctuation applies to systems through which energy and matter flow and whose macroscopic variables obey nonlinear equations. An ecosystem described by the equations of population dynamics corresponds to just such a situation, and it may now be proposed that biological evolution through mutation and selective advantage is yet another example of the principle of order through fluctuation (Allen, 1975, 1976). The equations of population dynamics describe the change of average genotype densities as a result of births and deaths of individuals, while it is supposed that the appearances of new genotypes as a result of spontaneous mutation play the role of fluctuations and are rare events compared to normal births.

In real systems, population densities tend asymptotically toward *stable* solutions, because the density fluctuations inherent in the statistical description will ensure that any solution which is not stable, that is, does not possess a "restoring force," will not be maintained. We assume therefore that just before a mutation occurs, the population densities have either attained a *stable* steady state or are described by a stable limit cycle. The mutant population, initially very small, will only result in an evolutionary step if its presence compromises the stability of the previous state.

We assume a general equation of change for the population densities; it takes the form

$$\frac{dX_i}{dt} = F_i(X_1, \ldots, X_n, X_{n+1}, \ldots, X_{n+\Delta}),\qquad i = 1, 2, \ldots, n, n+1, \ldots, n + \Delta \tag{A.1}$$

in which the genotypes existing before the occurrence of the mutation are $X_1, X_2, \ldots, X_n$ and the mutants are $X_{n+1}, X_{n+2}, \ldots, X_{n+\Delta}$. (A mutant allele may result in more than one new genotype.) The stability of Eq. (A.1) is found by solving

ISBN 0-201-03438-7/0-201-03439-5pbk.

$$\det |A_{ij} - \delta_{ij}\lambda| = 0 \tag{A.2}$$

where $A_{ij} = \partial F_i/\partial X_j$ at the state $X_1^0, X_2^0, \ldots, X_n^0$ and $X_{n+1} = X_{n+2} \ldots = X_{n+\Delta} = 0$. However, the terms $\partial F_i/\partial X_j$ where $i = n+1, \ldots, n+\Delta$ and $j = 1, 2, \ldots, n$ are zero because we have excluded, by the way we constructed our theory, the steady production of the mutants by the preexisting types. Thus, there can be no term in the mutant density equations which depends *only* on the genotypes $X_1, X_2, \ldots, X_n$. Inserting this result into (A.2), we have

$$\det|A_{ij} - \delta_{ij}\lambda| = \det|A_{k\ell} - \delta_{k\ell}\lambda| \times \det|A_{pq} - \delta_{pq}\lambda| = 0 \tag{A.3}$$

where $i,j = 1,2,\ldots,n+\Delta$; $k, \ell = 1,2,\ldots,n$; $p, q = n+1, n+2, \ldots, n+\Delta$.

The first determinant on the right-hand side, however, evaluated at the state $X_1^0, X_2^0, \ldots, X_n^0$; $X_{n+1} = X_{n+2} = \ldots = X_{n+\Delta} = 0$, gives the stability of the original system *before* the mutation occurred—and this, we assumed, was stable. Thus, if the mutant populations are to be amplified from zero, the second determinant on the right-hand side must possess a root with a positive real part. Thus, we can now write a criterion for the occurrence of an evolutionary step:

$$\det|A_{pq} - \delta_{pq}\lambda| = 0, \qquad p,q = n+1, n+2, \ldots, n+\Delta \tag{A.4}$$

must have a root λ with a positive real part when evaluated at the state $X_1^0, X_2^0, \ldots, X_n^0$ and $X_{n+1} = X_{n+2} = \ldots = X_{n+\Delta} = 0$.

Let us briefly illustrate the application of this result to a predator-prey ecosystem. We shall not treat explicitly the case of sexual reproduction, which has been discussed elsewhere and serves only to complicate the equations without changing the result in a qualitative way.

Let us consider a prey X_1, with birth rate k_1, which in the absence of the predator would grow logistically to a density N, and a predator Y with death rate d. The interaction between the two is characterized by $s_1 X_1 Y$:

$$\frac{dX}{dt} = k_1 X_1 (N - X_1) - s_1 X_1 Y, \qquad \frac{dY}{dt} = -dY + s_1 X_1 Y$$

This goes to the stable steady state, $X_1^0 = d/s_1$; $Y_1^0 = k_1/s_1\,(N - d/s_1)$. Let us suppose that a small quantity of a new prey type X_2 appears in our system. Our criterion gives us the condition that X_2 must fulfill if it is not to be rejected by the system. We have

ISBN 0-201-03438-7/0-201-03439-5pbk.

$$\frac{dX_2}{dt} = k_2X_2(N - X_1^0 - X_2) - s_2X_2Y$$

and hence

$$k_2(N - X_1^0 - X_2) - k_2X_2 - s_2Y^0 - \lambda = 0$$

Substituting into this $X_1^0 = d/s_1$; $Y^0 = k_1/s_1\ (N - d/s_1)$; $X_2 = 0$; we find that λ will be positive if

$$\left(N - \frac{d}{s_1}\right)\ \left(k_2 - \frac{s_2k_1}{s_1}\right) > 0$$

Since $N - d/s_1 > 0$, we must have

$$\frac{k_2}{s_2} > \frac{k_1}{s_1} \tag{A.6}$$

Thus, only mutants fulfilling this condition can be amplified by the system and a prey evolution alone will lead to a steady increase in k/s.

A similar analysis for the predators shows that d/s will decrease with each evolutionary step. Taking the evolution of both predator and prey, we see that the coefficient s has no well-defined direction of drift but that k increases and d decreases. This tells us that the expression

$$\frac{Y}{X} = \frac{k}{d}\left(N - \frac{d}{s}\right)$$

which is the ratio of predator to prey at a given steady state, will increase as evolution proceeds.

Another very recent application of the criterion described above considers species with a choice of a number of different resources which biological evolution could lead them to exploit. The criterion leads to the prediction that a rich environment, where each resource is available in large quantities, will cause species to evolve into "specialists," whereas a poor environment, where each type of resource is only present in small quantities, will lead to "generalists." In order to test this prediction, the distribution of finches on the Galapagos Islands was studied, and the prediction is borne out by the number of finch species occupying identical vegetation zones on large and very small islands. Other applications are under active consideration.

ISBN 0-201-03438-7/0-201-03439-5pbk.

The sources of "innovation" need not necessarily be genetic, but may also refer to changes in behavior in a species with imitative mechanisms. The adoption of new techniques by means of this type of evolution does not require the destruction of the less adapted types and thus represents a possibly faster channel of evolution. Associated with this mode is the tendency to evolve cooperative groups, characterized by the division of labor, hierarchical relationships and "castes," as well as by mechanisms of population regulation, and even by altruism. For example, we find that division of labor and castes appears in insect societies as the result of evolution of large colonies existing in a rich medium. The competitive unit, subject to selection, in this case is not the individual, but the *group* (see also Chapter 1 by Waddington). The possibility of its existence and of cooperative mechanisms coming into play depends to a large extent on the transport properties and communication channels of the medium. In Sections 5 and 6, the use of chemotaxis by amoebas and insect societies was discussed, but groups of higher animals may use a large variety of techniques ranging from visual and audible to chemical signals. The use of language in the human domain signifies a further decisive step in this direction.

The application of the principle of order through fluctuation to ecosystems of interacting populations leads to a criterion that must be fulfilled for evolution to occur, and hence to the possibility of predicting, in certain configurations, the long-term thrust of biological evolution.

REFERENCES

Allen, P. M. (1975). "Darwinian Evolution and a Predator–Prey Ecology," *Bull. Math. Biol.*, **37**, 389–405.

Allen, P. M. (1976). "Evolution, Population Dynamics and Stability," *Proc. Natl. Acad. Sci.*, **73**, 665–668.

Althusser, L., and Balibar, E. (1973). "L'objet du *Capital.*" In *Lire le Capital* (F. Maspero, ed.), Vol. II, p. 61. Paris: Maspero.

Andronov, A., Vitha, A., and Chaikin, S. (1966). *Theory of Oscillations* (Engl. transl. by S. Lefschetz). Princeton, N.J.: Princeton Univ. Press.

Aron, R. (1967). *Les étapes de la pensée sociologique.* Paris: Gallimard.

Auchmuty, J. F. G., and Nicolis, G. (1975). "Bifurcation Analysis of Non-Linear Reaction–Diffusion Equations," *Bull. Math. Biol.*, **37**, 323.

Auchmuty, J. F. G., and Nicolis, G. (1976). "Bifurcation Analysis of Non-Linear Reaction–Diffusion Equations, II: Chemical Oscillations," *Bull. Math. Biol.*, in the press.

ISBN 0-201-03438-7/0-201-03439-5pbk.

Babloyantz, A. (1972). "Far from Equilibrium Synthesis of 'Prebiotic Polymers'," *Biopolymers*, **11**, 2349–2356.

Babloyantz, A., and Hiernaux, J. (1975). "Models for Cell Differentiation and Generation of Polarity in Diffusion Governed Morphogenetic Fields," *Bull. Math. Biol.*, **37**, 637–657

Bergson, H. (1963). *Evolution Créatrice*, Eds. du Centennaire. Paris: Presses Universitaires de France.

Boiteux, A., and Hess, B. (1974). "Oscillations in Glycolysis, Cellular Respiration and Communication." In *Physical Chemistry of Oscillatory Phenomena* (Faraday Symposium 9). London: Faraday Division of The Chemical Society.

Boltzmann, L. (1872). "Weitere Studien über das Wärmegleichgewicht unter Gasmolekülen," *Ber. Akad. Wiss. Wien*, **66**, 275.

Charbonnier, G. (1969). *Conversations with Claude Lévi-Strauss.* (T. and D. Weightman, transl.) London: Jonathan Cape.

Clausius, R. (1857). "Über die Art der Bewegung welche wir Wärme nennen," *Ann. Physik*, **100**, 353.

Deneubourg, J. L., and Nicolis, G. (1976). *Proc. Natl. Acad. Sci.*, in the press.

Durkheim, E. (1973). *De la Division du Travail Social.* Paris: Presses Universitaires de France.

Eigen, M. (1971). "Self-Organization of Matter and the Evolution of Biological Macromolecules," *Naturwiss.*, **58**, 465–522.

Erneux, T., and Herschkowitz-Kaufman, M. (1975). "Dissipative Structures in Two Dimensions," *Biophys. Chem.*, **3**, 4, 345.

Faraday Symposium 9 (1974). *Physical Chemistry of Oscillatory Phenomena.* London: Faraday Division of The Chemical Society.

Gallais-Hamonno, F., and Chauvin, R. (1972). "Simulation sur ordinateur de la construction du dôme et du ramassage des brindilles chez une fourmi *(Formica polyctena)*," *C. R. Acad. Sci. Paris*, **275D**, 1275.

Gerisch, G., and Hess, B. (1974). "Cyclic-AMP-Controlled Oscillations in Suspended Dictyostelium Cells: Their Relation to Morphogenetic Cell Interactions," *Proc. Natl. Acad. Sci.*, **71**, 2118.

Glansdorff, P., and Prigogine, I. (1971). *Thermodynamic Theory of Structure, Stability, and Fluctuations.* New York: Wiley-Interscience.

Goël, N. S., Maitra, S. C., and Montroll, E. W. (1971). "On the Volterra and Other Nonlinear Models of Interacting Populations," *Rev. Mod. Phys.*, **43**, 231.

Goldbeter, A. (1973). "Organisation spatio-temporelle dans les systèmes enzymatiques ouverts," Thèse de doctorat, Université Libre de Bruxelles.

Goldbeter, A. (1975). "Mechanism for oscillatory synthesis of cyclic AMP in *Dictyostelium Discoideum*," *Nature*, **253**, 540.

Grassé, P. P. (1959). "La reconstruction du nid et les coordinations interindividuelles chez *Bellicositermes natalensis* et *Cubitermes sp.* La théorie de la stigmergie: essai d'interprétation du comportement des termites constructeurs," *Insectes Sociaux*, **6**, 41–83.

Gusdorf, G. (1971). *Les Principes de la Pensée au Siècle des Lumières.* Paris: Payot.

ISBN 0-201-03438-7/0-201-03439-5pbk.

Hangartner, W. (1967). "Spezifität und Inaktivierung des Spurpheromons von Losius fuliginosus Latr. und Orientierung der Arbeiterinnen im Duftfeld" *Z. Vergleichende Physiol.* **57** (2), 103.

Hanson, M. P. (1974). "Spatial Structures in Dissipative Systems," *J. Chem. Phys.*, **60**, 3210–3214.

Herschkowitz-Kaufman, M. (1973). "Quelques aspects du comportement des systèmes chimiques ouverts loin de l'équilibre thermodynamique," Thèse de doctorat, Université Libre de Bruxelles.

Herschkowitz-Kaufman, M., and Nicolis, G. (1972). "Localized Spatial Structures and Non-Linear Chemical Waves in Dissipative Systems," *J. Chem. Phys.*, **56**, 1890–1895.

Janne, H. (1963). *Le Système Social: Essai de Théorie Générale.* Brussels: Editions de l'Institut de Sociologie de l'Université Libre de Bruxelles.

Keller, E., and Segel, L. A. V. (1970). "Initiation of Slime Mold Aggregation Viewed as an Instability," *J. Theoret. Biol.*, **26**, 399.

Lefever, R. (1968). "Stabilité des structures dissipatives," *Bull. Classe Sci. Acad. Roy. Belg.*, **54**, 712.

Lotka, A. J. (1956). *Elements of Mathematical Biology.* New York: Dover.

Malek-Mansour, M., and Nicolis, G. (1975). "A Master Equation Description of Local Fluctuation," *J. Stat. Phys.*, **13**, 197.

May, R. (1973). *Model Ecosystems.* Princeton, N.J.: Princeton Univ. Press.

Mazo, R. (1975). "On the Discrepancy between Results of Nicolis and Saito concerning Fluctuations in Chemical Reactions," *J. Chem. Phys.*, **62**, 10, 4244.

McQuarrie, D. A. (1967). *Supplementary Review Series in Applied Probability.* London: Methuen.

Nazarea, A. D. (1974). "Critical Length of the Transport-Dominated Region for Oscillating Non-Linear Reactive Processes," *Proc. Natl. Acad. Sci.*, **71**, 3751.

Nicolis, G., and Auchmuty, J. F. G. (1974). "Dissipative Structures, Catastrophes, and Pattern Formation: A Bifurcation Analysis," *Proc. Natl. Adad. Sci.*, **71**, 2748.

Nicolis, G., and Babloyantz, A. (1969). "Fluctuations in Open Systems," *J. Chem. Phys.*, **51**, 6, 2632.

Nicolis, G., and Prigogine, I. (1971). "Fluctuations in Non-Equilibrium Systems," *Proc. Natl. Acad. Sci.*, **68**, 2102–2107.

Nicolis, G., and Prigogine, I. (1974). "Thermodynamic Aspects of Spatio-Temporal Dissipative Structures." In *Physical Chemistry of Oscillatory Phenomena* (Faraday Symposium 9). London: Faraday Division of The Chemical Society.

Nicolis, G., and Prigogine, I. (1976). *Self-Organization in Nonequilibrium Systems: From Dissipative Structures to Order through Fluctuations.* New York: Wiley-Interscience (in the press).

Nicolis, G., Malek-Mansour, M., Kitahara, K., and Van Nypelseer, A. (1974). "The Onset of Instabilities in Nonequilibrium Systems," *Phys. Letters*, **48A**, 217.

Nitzon, A., Ortoleva, P., and Ross, J. (1974). "Stochastic Theory of Metastable Steady-State and Nucleations." In *Physical Chemistry of Oscillatory Phenomena* (Faraday Symposium 9). London: Faraday Division of The Chemical Society.

ISBN 0-201-03438-7/0-201-03439-5pbk.

Onsager, L. (1931). "Reciprocal Relations in Irreversible Processes, I," *Phys. Rev.*, **37**, 405.

Pacault, A., de Kepper, P., Hanusse, P., and Rossi, A. (1975). "Etude d'une réaction chimique périodique: Diagramme des états," *C. R. Acad. Sci. Paris*, **281C**, 215.

Perroux, F. (1964). *Industrie et Création Collective*, Tome I. Paris: Presses Universitaires de France.

Portnow, J. (1974). Discussion Remarks. In *Physical Chemistry of Oscillatory Phenomena* (Faraday Symposium 9). London: Faraday Division of The Chemical Society.

Prigogine, I. (1967). *Thermodynamics of Irreversible Processes*, 3d ed. New York: Wiley-Interscience.

Prigogine, I. (1972). "La Thermodynamique de la Vie," *La Recherche*, **3**, 547.

Prigogine, I. (1973). *Physique et Métaphysique*, lecture given at Académie Royale de Belgique on the occasion of its bicentenary.

Prigogine, I., and Herman, R. (1971). *Kinetic Theory of Vehicular Traffic.* New York: American Elsevier.

Prigogine, I., Nicolis, G., and Babloyantz, A. (1972). "Thermodynamics of Evolution," *Phys. Today*, **25**, Nos. 11 and 12.

Prigogine, I., Nicolis, G., Herman, R., and Lam, T. (1975). "Stability, Fluctuations, and Complexity," *Cooperative Phenomena*, **2**, 103-109.

Rachet, G. (1969). *Archéologie de la Grèce préhistorique, Troie-Mycène, Cnossos.* Verviers: Marabout Université.

Rettenmeyer, C. W., "Behavioral Studies of Army Ants," *Kansas Univ. Bull.*, **44**, 281.

Susman, M. (1964). *Growth and Development.* Englewood Cliffs, N.J.: Prentice-Hall.

Tarde, G. (1890). *Les Lois de l'Imitation.* Paris: Alcan.

Weidlich, W. (1974). "Dynamic of Interactions of Several Groups." In *Cooperative Phenomena* (H. Haken, ed.) Amsterdam: North-Holland.

Wilson, E. O. (1971). *The Insect Societies.* Cambridge, Mass.: Harvard Univ. Press.

Zotin, A. I. (1972). *Thermodynamic Aspects of a Developmental Biology.* Basel: S. Karger.

ISBN 0-201-03438-7/0-201-03439-5pbk.

6

Vibrations and the Realization of Form

Ralph Abraham

Universal Form and Harmony
were born of Cosmic Will,
and thence was Night born, and thence
the billowy ocean of Space;
and from the billowy ocean of space
was born Time—the year
ordaining days and nights,
the ruler of every movement.

Rigveda X 190

1. INTRODUCTION TO MACRODYNAMICS

Macrodynamics is a synonym for kymatics. My preference for Anthony's (1969) nomenclature over Jenny's (1967) is just personal taste. If any of this part seems too technical, skip directly to Section 2.

Morphogenesis, the evolution of form from chaos, has a high priority in the philosophical literature of many cultures: the *Rigveda, I Ching*, Heraclitus, Cabala, and others. Up to this very volume, phenomenological descriptions of morphogenesis in various spheres abound in our literature. On the other hand, *morphodynamics*—the study of the mechanics of morphogenetic processes in the context of hard science—is just beginning. It has been born of two recent developments: a suitable mathematical foundation, the *theory of catastrophes* of René Thom (1973); and an adequate observational tool, the *macroscope* of Hans Jenny (1967; 1972). Here, then, is a very concise introduction to *experimental morphodynamics*, including a preliminary report on our own macron observations through the first color macroscope. This chapter is dedicated to Neemkaroli Baba, late of Uttar Pradesh.

ISBN 0-201-03438-7/0-201-03439-5pbk.

Erich Jantsch and Conrad H. Waddington (eds.), *Evolution and Consciousness: Human Systems in Transition.*

Simple Macrons

Macrodynamic processes in nature take place in hierarchical systems of compound (heterogeneous) macron organisms. To understand these processes, we try to dissect them into fictitious categories of simple (homogeneous) macrons. The three basic categories are physical (P); chemical (C); and electrical (E). The physical macrons are further subdivided according to the material state of the macron medium: solid (PS), isotropic liquid (PL), liquid crystal (PX), and gas (PG). Here we discuss examples of these six types.

Physical Solid (PS): A flat plate is vibrated transversally by an external force, usually electromechanical transducers coupled either directly or through an intermediate fluid. A stable aspect of the system is a spiderweb of motionless curves, the Chladni nodal lines, originally observed by sprinkling sand on the plate. The complete vibration pattern of the plate is best revealed by laser interferometry. This pattern is the *macron* in this example. It depends upon control parameters of two types: *intrinsic controls*, such as dimensions and elasticity of the medium; and *extrinsic controls*, such as frequency and amplitude of the driving force. Of course, this example is very special, as the medium is more or less two dimensional. For a generic example in this category, consider a rubber ball in place of the thin plate. Stable modes of vibration are characterized by symmetric distortions of shape, separated by motionless nodal surfaces. If the medium is magnetic or piezoelectric, driving forces may be applied directly with electromagnetic fields.

Physical Isotropic Liquid (PL): Beginning once again with a two-dimensional approximation, suppose a round dish is filled with a thin layer of isotropic liquid, and the bottom of the dish is heated. Soon the liquid will begin to *simmer.* Careful observation will reveal a spiderweb of nodal lines (actually, parallel lines—*rolls* and packed hexagons called *Bénard cells*—are combined in patterns), within which the liquid convects toroidally (up at the boundary of the cell, down in the center). This *Bénard phenomenon* is a macron. Another type is observed by vibrating the bottom of the dish in a (PL) macron. As the amplitude is gradually increased, the liquid layer first behaves as a solid—the *elastic macron;* then, after a certain critical amplitude is reached, the *simmering point*, a convection or Bénard-type simmering fluid flow begins—the *hydrodynamic macron.* The macrons, or stable modes, depend on intrinsic controls such as shape, compressibility, and viscosity; and external controls such as frequency and amplitude of the driving force. In the general case of a thick layer of liquid, the elastic and hydrodynamic macrons are three-dimensional generalizations of these effects. But

ISBN 0-201-03438-7/0-201-03439-5pbk.

there occurs at least one effect of a different type. If the dish is rotated or the liquid is stirred, there may arise toroidal partitions, within which a ring of fluid—a *Taylor cell*—flows spirally. These rings are also seen when a drop of fluid enters another mass of fluid, as in smoke rings. Hierarchical repetition of Taylor cells may be observed by dripping ink into a glass of very still water. The *von Karman vortex street* also belongs to this class.

Finally, we include in this category *isotropic powder*, that is, dust made of spherical solid particles of identical size. Jenny (1967; 1972) has produced Bénard cells in powders of moss spores.

Physical Liquid Crystal (PX): If the medium is in a liquid crystal metaphase, any macrons of elastic or hydrodynamical type may be induced in it. But two additional phenomena have been observed which are peculiar to this phase, and other simple macrons will undoubtedly be discovered which belong especially to this category. If a thin layer of fluid is exposed to a transverse electrostatic field, simmering is induced in approximately hexagonal cells—the *Williams effect.* Presumably, an elastic macron is induced below the simmer point. In an oscillating electromagnetic field, piezoelectric waves are induced—the *flexoelectric effect.* In this category we might also include *anisotropic powder*—dust of identical aspheric solid particles.

Physical Gaseous (PG): In gases, we observe the macrons of isotropic liquid, as well as (presumably) additional pattern mechanisms belonging specifically to this category. Perhaps these gaseous macrons are unique combinations of elastic and hydrodynamical macrons of the (PL) category, possible in this context because of the high compressibility and low viscosity of the usual gases. The enormous dimensions of these macrons make them hard to observe, and at present it is not known whether or not exclusively gaseous macrons exist.

These four classes of macrons compose the category (P) of simple physical macrons. This is the context of most of the research in experimental morphology up to now. The remaining two categories, (C) and (E), are therefore very embryonic at present.

Chemical (C): There are various macrons, or basic pattern phenomena, which are fundamentally chemical in origin. These occur in heterogeneous media, amid chemical reactions. Included are mechanisms of change of state, such as patterns of precipitation, Liesegang rings of crystallization, and opalescences like abalone shell. To this class also belong the classical diffusion patterns as well as the newly discovered patterns of periodic chemical reactions (see Chapter 5 by Prigogine). This is a little-studied category, which will undoubtedly be explored more thoroughly.

ISBN 0-201-03438-7/0-201-03439-5pbk.

Electrical (E): The description of basic electrical macrons is included here for the sake of completeness—in spite of being based almost entirely on speculation—and because of my belief that it will figure vitally in the understanding of the brain and in the engineering of artificial intelligence, sometime in the future.

Consider a heterogeneous medium of smoothly changing physical properties, especially electrical conductivity, and possibly containing sources of charge. This is an *electronic spacework.* An electronic network may be thought of as a spacework with discontinuities, or as a retraction of a spacework onto its skeleton of dimension one. A semiconductor device is an example of a genuinely three-dimensional spacework. However, this concept must be allowed to include matter in all phases, especially charged fluid (plasma, ionized gas, etc.). Thus, classical magnetohydrodynamics (MHD) is included in this context.

In an electronic spacework, subject to controlled external electromagnetic fields or to controlled charge exchange with the environment, macrophenomena of categories (P) or (C), as well as other unique phenomena, may be observed. Those macrons occurring uniquely in the context of spaceworks include category (E). For example, the Störmer orbits and Alfvén waves of magnetogasdynamics and northern lights are macrons of type (E). As in ionized fluids, rolls comprise transformers, Bénard cells are toroidal inductors, Taylor cells combine linear and toroidal induction, membranes are capacitors, and so forth. It may be expected that an entirely new discipline of engineering could be based upon a full understanding of macrons in specific spaceworks. The idea of a liquid crystal transistor, combining fluid, electronic, and MHD technologies, is not too far-fetched.

Macrons of type (E) will be known better in the future, when the development of specific MHD machines will make systematic observation possible.

Complex Systems of Macrons

The macrodynamics of a real event is complex in two ways. First, a single organic structure may exhibit a macron in which physical, chemical, and electrical modes are combined. This is especially the case with biological organisms. Second, two distinct structures may be weakly coupled, forming a larger, compound organic unit. Here we discuss compound modes and coupling separately.

Since basic macrons are of three types, (P), (C), and (E), there are only four types of compound macrons: (P–C), (P–E), (C–E), and (P–C–E).

ISBN 0-201-03438-7/0-201-03439-5pbk.

Physical-Chemical (P–C): A typical situation of this type is a mixture of fluid reagents. While a stable pattern of chemical origin exists, an elastic or hydrodynamic macron is excited. Since convection is faster than diffusion, the hydrodynamic macron dominates the patterns of reagent concentration and reaction rate. For example, Lew Howard and Nancy Kopell (1976) observed Bénard cells with purple hexagonal boundaries and red central cell bodies in the Zhabotinsky reaction, when the surface of the fluid was cooled by evaporation. The separation of the reagents is accomplished by a *separation mechanism*, which, for this *Howard-Kopell phenomenon*, is undoubtedly centrifugation of the reagents. The separation mechanism is also well illustrated by an analogous experiment carried out by Jenny (1967; 1972): Sand is sprinkled on a vibrating plate; it gravitates, very slowly, to the Chladni nodal lines. These motionless curves outline cells of transverse vibration, each with a center, or nucleus, of maximum motion. Now spore powder is sprinkled on the vibrating plate. This moves to the nuclei, forming small piles of powder at each nucleus. Furthermore, each pile can be seen to roll constantly in a toroidal eddy, exactly as in a Bénard cell. In the latter case, I believe the separation mechanism is differential response of the reagents to flotation in invisible Bénard cells excited in the air over the plate by the vibration.

Physical-Electrical (P–E): As in the previous discussion of basic macrons of type (E), we can only speculate on this case, which is exemplified by *plasma.* The production of a toroidal inductor in a fluid spacework by intentional excitation of a Bénard cell is an example of a compound (P–E) macron. The generation of electromagnetic waves by physical vibration of a cholesteric flexoelectric liquid crystal is another.

Chemical-Electrical (C–E): Spaceworks designed specifically to separate ionized components into a particular spatial pattern could be used to grow semiconductor crystals, specialized lenses, or any frozen, precipitated, or crystallized solid in a given pattern. These media are *electrochemical spaceworks.*

Physical-Chemical-Electrical (P–C–E): Physical macrons in fluid reagent mixtures, including liquid crystal and solid components, some of which are charged or otherwise electroactive, comprise the patterns of living organisms. Embryology provides countless examples. This general case may therefore also be called *bioplasma.*

Whereas in the laboratory, macrons of a pure, basic type can be created, in the real world of phenomena only the general case is found. Suppose now that two single systems of bioplasmic (P–C–E) type are at hand, and their separate stable modes are known. Let these two now be weakly coupled, by physical contact, chemical mixing in

ISBN 0-201-03438-7/0-201-03439-5pbk.

small exchanges, interaction through the electromagnetic field, or a combination of these means. The coupled system will now have its own stable modes. How are the combined macrons related to the original separate macrons? This relation, which we call the *algebra of macrons*, is the most intriguing problem of morphology. The classical ideas of resonance, sympathetic vibration, and so on, serve as clues. The experiment of Jenny (1967; 1972), showing the vibrating plate and the overlying air in related macrons (revealed by the eddying piles of powder at the nuclei of the plate macron), gives a more useful example of macron addition.

In fact, there are no pure macrons. In order to study the stable modes of one system, we must couple it to another. Thus, all experiments in morphology are actually examples of coupled macrons, and because the number of organic units in a coupled system of the phenomenal universe is always large, we shall one day be led to a probabilistic, or statistical-mechanical, theory of complex macrons for hierarchical systems. However, this is far off at the moment. We have, at present, only a very rudimentary preview of the mathematical theory of basic macrons.

Geometry of Macrons.

A full understanding of the mathematical description of macrons would require a knowledge of the theory of dynamical systems up to the current research frontier and beyond. For those who wish to pursue this exciting hobby, the introductory book of Hirsch and Smale (1974) provides a good starting point. For our presentation, we shall require but a single concept of that theory, that of *attractor*, which is easily grasped on an intuitive level.

Suppose that a particular medium is to be studied, for example, a bowl of salty jelly. We have to assume (1) that a suitable mathematical space has been described, called the *phase space*, such that each point in the phase space corresponds to a completely satisfactory description of a configuration, or geometrical posture, of the jelly; and conversely, that each posture of the jelly corresponds to a unique labeling point in the phase space. Therefore, jiggling the jelly defines a curve: a point in the phase space, moving along a path. Next, we have to assume (2) that the particular experimental situation of the jelly—for example, if it is stirred in a precise way—is described by a *dynamical system* in the phase space. This is a mathematical structure with the following properties:

ISBN 0-201-03438-7/0-201-03439-5pbk.

(a) The phase space is divided into a number of different zones, called *basins of attraction* (see also Chapter 4 by Holling, Section 3);
(b) In each basin there is a distinguished set, its *attractor*, which is a sort of atomic, or basic, representation of a dynamical system;
(c) If the jelly, in the chosen experimental situation, is set going in any original state, its corresponding curve of successive states in the phase space will proceed in a unique fashion toward a final equilibrium motion near the attractor of the basin in which the curve started.

If we ignore here the nonequilibrium states—which is justified for structures for which the approach to equilibrium is very swift—then all we need to know about dynamical systems is their attractors, which is excellent, because all dynamical systems may then be represented by the same types of "atomic" attractors, in different "molecular" clusters. Further, these common attractors are classified by a system which begins with a simple sequence, starting with the simplest. Here are the first three attractors in this sequence:

1. a single point, corresponding to static equilibrium—the jelly ceases to jiggle, or dies;
2. a circle, with a parameter, corresponding to a cyclic repetition of states, an oscillation in the jelly with a single period, or frequency;
3. a two-dimensional torus, with a curve spiraling indefinitely around it, as in toroidal inductors—corresponding to an *almost periodic motion*, a compound oscillation with two independent frequencies, irrationally related.

This sequence continues with tori of increasing dimension and more complicated compound oscillations, until very chaotic motions are included. The full description of this list of attractors is, in my view, one of the great achievements of mathematics in the twentieth century.

This completes our excursion into dynamical systems theory, and its concept of attractor. By now the intention of this excursion is probably clear: *the mathematical description of a macron is an attractor.* In itself, this does not help us much to understand macrodynamics or morphogenesis.[1] It is actually the theory of *transitions of attractors*, or *catastrophes*, as developed by René Thom (1973), which is the basis for the geometry of macrons, as it has developed so far. Here is the concept of catastrophe, as used in dynamical systems theory.

Returning to our original experimental situation, we have supposed that the medium of the experiment—salty jelly (or an economy, an

[1] It should be noted that "atomic" attractors, at this stage of theory building, represent global stability and therefore cannot, by themselves, adequately describe nonlinear behavior, in particular, the amplification of fluctuations which may drive the system to the point of a catastrophe, or qualitative change from one attractor to another. (Comment by E. J., editor)

ISBN 0-201-03438-7/0-201-03439-5pbk.

electronic black box, or whatever)—is described by a phase space; and that the specific experimental situation—comprising the fixed values of the various intrinsic and extrinsic control parameters—is described by a particular dynamical system, with its molecular cluster of basins and attractors. It follows, then, that changing the values of the controls will change the dynamical system, the basins, and the attractors, which correspond to observable states, or macrons. Thus, as the controls are smoothly changed, the attractor under observation must be expected to change. But the attractors belong to a discrete list, and can only change in jerks. These are the catastrophes and they may be considered as boundaries of particles, providing a mathematical (nonlinear spectral theory) version of particle–wave duality. In fact, an esoteric quantum field theory based on this analogy has been suggested by Thom, and may one day be developed.

Dynamical systems theory provides us, in addition to the classification of attractors—point, circle, torus, and so forth—, with a classification (not yet complete) of catastrophes, or allowable (i.e., generic) transitions of attractors. This classification also begins with a discrete list of increasingly complex phenomena. We end this mathematical aside with a description of the two simplest types of catastrophes: the *leap* and the *wobble*.

Suppose the system is observed in a certain macron (attractor) and the control parameters are gradually changed. The macron is gradually distorted, but undergoes no definite change of type. Suddenly, at a critical value of the controls, it changes instantaneously into a completely different macron. This is a *leap*. In the simplest case, it is a point (steady-state) attractor which leaps. The steady state suddenly changes to a radically different steady state. For example, the onset of Taylor cells in rotating fluids is a leap catastrophe.

The wobble is a subtle catastrophe, almost unnoticeable. In the simplest case, called *Hopf excitation*, a point attractor changes into a circular attractor, as the control parameters are changed through a critical value. At first, the circle is very small, corresponding to a *wobble*, or oscillation of very small amplitude. As the controls continue to change, the circle (and wobble) grows until the oscillation becomes noticeable. For example, the fluttering of the boundaries of Taylor cells in rotating fluids is an example of a wobble catastrophe.

It is possible to organize all the attractors and catastrophes, referring to a given experiment with controls, in a single geometric model. This is called the *logos*. The structure of these models, the *geometry of macrons*, is the triumph of Thom's theory of catastrophes. Unfortunately, it is unapproachable without technicalities. In an experimental situation, however, it is possible to construct a map of the logos more or less empirically by exploration. The confusing feature is a kind of

ISBN 0-201-03438-7/0-201-03439-5pbk.

hysteresis: the observed state, for a given control setting, may depend on the direction of approach to that control setting. For each control value (such as rate of stirring of the salty jelly), many attractors may exist. The geometry of macrons can be a great help to the experimenter at this point. This will become clearer in the context of the example discussed in the next section.

For those who would like to know more of macron geometry, the basic references are included in the bibliography. My "Introduction to Morphology" (R. Abraham, 1972) includes some helpful illustrations. *Warning:* The classical literature of catastrophe theory (Thom, 1973, and Zeeman, 1971) assumes that the phase space is finite dimensional. This is not the case in macron theory. The extension to the infinite-dimensional case, technically very difficult and not completely satisfactory, is discussed by Ruelle and Takens (1971) and by Marsden and McCracken (1976).

Techniques of Macroscopy.

The study of macrodynamics must be founded on the observation of basic macrons of types (P), (C), and (E), and their coupling behavior. Here we describe the construction and operation of the *macroscope*, a universal tool for the observation of transparent macrons of physical (P) types based upon prototype instruments built by von Békésy (1960), Jenny (1967), Schwenk, Settles (1971), and others.

The instrument combines five units (see Figure 6.1): (1) a color schlieren-optical system, of Settles–Toeplitz type, with a four-inch field of view, terminating in a rear projection screen; (2) a transparent vibrating dish, driven by a high-fidelity loudspeaker outside the field of view; (3) a sine-wave generator, controllable in the rectangle: 0–1000 Hertz by 0–15 watts; (4) a control rectangle monitor, including cathode-ray tube and two digital meters; and (5) a xenon arc lamp, capable of microsecond flashes up to 1000 Hertz at 100 watts average power, triggered by (i.e., synchronous with) the sine-wave generator, with adjustible phase lag.

In operation, the fluid or elastic medium (which must be perfectly transparent) is placed in the transparent dish. The instrument is switched on, and the experimenter steers the control parameter around the rectangle with two knobs, while watching the colored image on the screen. Leap and wobble catastrophes are readily observed, and can be plotted on the face of the CRT control monitor with a wax crayon. The geometry of the logos is easily discovered by exploration. The exploration of different media indicates the effect of the intrinsic control

ISBN 0-201-03438-7/0-201-03439-5pbk.

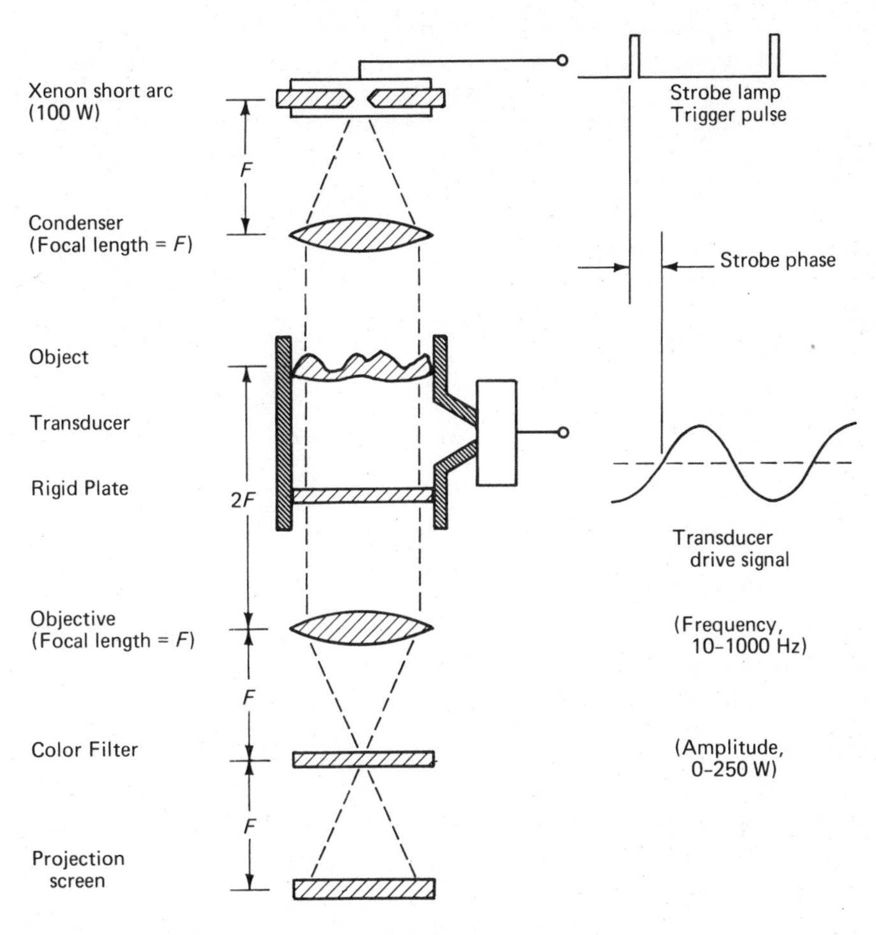

FIGURE 6.1. Schematic view of the four-inch macroscope of the University of California, Santa Cruz. (Diameter, 4 in.; F, 48 in.)

parameters—for example, physical dimensions and viscosity—upon the logos.

But what is the relationship between the colored image on the screen, the physical macron within the medium, and the mathematical attractor which describes it? Theoretically, the physical parameter represented on the screen is the *horizontal gradient vector field of the index of refraction of the medium, expressed in polar coordinates of color and intensity.* In practice, interpretation in macroscopy, as in radiology, is learned by experience. Two separate causes of coloring must be distinguished: deformation of the surface of the medium (lenticulation) and pressure waves within it (pressurization). Normally, two separate images are superimposed on the screen, the λ-image (due to

ISBN 0-201-03438-7/0-201-03439-5pbk.

lenticulation) and the π-image (due to pressurization). Fluid flow within the medium is not revealed, but can be observed directly by the usual technique: dusting the medium with aluminum powder. Also, rapid Bénard cells (boiling) cause concentric rings in the λ-image, and discs in the π-image.

The image, with the control parameters left fixed, is usually moving. In fact, it is full of fast action (e.g., like boiling) and also presents a slow progression through different forms.[2] The slow motion repeats itself periodically. This is a toroidal attractor. Counting the dimensions of the torus strains the human space-time pattern recognition facility, and justifies the warning of Anthony (1969) that macroscopy causes brain damage. But when the driving signal is very small, the image may be still. This does not mean that the macron is a point attractor (stable equilibrium), because the illumination is stroboscopic, and stops all periodic motion at the driving frequency. At this point, the phase between the driving signal and the arc lamp must be adjusted through a full cycle to determine whether the macron is a point or circle attractor.

Macroscopy is impossible to describe verbally or photographically. Color cinematography and videotape cassette are the appropriate media for registration of experimental data in this field, and in experimental morphodynamics in general. Moreover, by using your imagination freely, you may think of countless experiments to do with a macroscope, the results to be stored in color videotape cassettes. Also, many different macroscopic devices are feasible, including one under development at present, in which video equipment itself is used as an analogue device to generate macrons and catastrophes.

2. APPLICATIONS TO MORPHOGENESIS

A long series of applications of catastrophe theory to morphogenesis already exists, thanks to the inspired works of Thom (1973) and Zeeman (1971; see also Isnard and Zeeman, 1975). The majority of these applications belongs to static theory and shows that the geometry of point macrons alone is adequate to model a fantastic variety of morphological phenomena in the real world. Therefore, in this section, I shall give a selection of sample applications which are essentially nonstatic, or vibratory, in nature. These are from the traditional four levels of the phenomenal universe.

[2] The cover of this volume shows the nucleation of a new macron, photographed by the author with the macroscope of the University of California, Santa Cruz.

ISBN 0-201-03438-7/0-201-03439-5pbk.

Cosmology.

There is not much to say on this level beyond the basic observation of Jenny (1967; 1972): sand patterns on vibrating plates are analogous to galactic patterns of stellar material. If this analogy is pursued further, there arises a classical conundrum: What cosmic driving force corresponds to the plate, and how is it coupled to the galactic dust? This is the basic problem of the priority of the *word* in the philosophy of the Cabala, or the *tapas* of the *Rigveda*, which I have transliterated as Cosmic Will in the preface. In any case, it is beyond mathematics, I think.

Geology.

Here I can cite a few sample applications from each of the three basic planetary spheres. Regarding the morphogenesis of the *geosphere*, a basic morphogenetic situation is presented by the condensing sequence of gaseous, liquid, and solid phases, which could be studied in the macroscope. I suppose the conservation of the vorticity inherited from the initial motion of the cosmic material determines a certain macron in the sphere of mixed phases, combining elastic lenticulation of the crust—determining the location of continents, floating mountain ranges, ocean basins, and perhaps a network of global rifts along the nodal surfaces—with Bénard cells of convection in the hotter liquid core. These cells may be the driving force of continental drifts and earthquakes.

In the *hydrosphere*, I suspect that global ocean currents are toroids of the Taylor cell type. Local temperature gradients must produce Bénard cells, some of which may be very stable. Perhaps these are responsible for sculpturing the conical projections of the ocean floor. On a smaller scale, Bénard cells are obviously responsible for the honeycomb patterns on the bottom of icebergs observed by the Jacques Cousteau group.

Macrons in the *atmosphere* are manifest in the wind patterns of the weather map. Bénard cells cause honeycombs in sand dunes and sun cups on glaciers. It is not unlikely that the prevailing westerly winds contain Taylor cells girdling the equator. Hydrodynamical macrons around spinning spheres probably deserve closer study. The macroscope is an ideal tool for such practical investigations.

Biology and Neurophysiology.

Much has been written on biological morphogenesis (see Waddington, 1968–1972) and undoubtedly there is much more to come. The book of d'Arcy Thompson (1945) has become a modern classic. Comparison of

ISBN 0-201-03438-7/0-201-03439-5pbk.

the nature drawings of Haeckel (1974), or of the photographs of Strache, with the macrophotos of Jenny (1967; 1972) is very suggestive. Turing's (1952) revolutionary article on phylotaxis was perhaps the starting point of modern mathematical morphogenesis. The mechanisms of chemotaxis and ecotaxis are active areas of research. Macrodynamic explanations of nongenetic heredity, orgasm, telepathy, and many other phenomena are easily proposed. I shall confine myself here to two applications which are the subject of current empirical study: the ear and the brain.

The process of audition is more or less understood, except for the mechanical to neural transducer, the cochlea. This is a closed vessel of fluid (perilymph) with a mechanical input piston on one end, and a very complex pressure-sensitive organ stretched within the fluid and comprising a flection sensor (organ of Corti) embedded in a jelly (endolymph) bound by two membranes (Reissner and basilar). Obviously, this is a natural macroscope. Realizing this, von Békésy (1960), the great pioneer of perception research, made a transparent model of the cochlea and looked at the macron produced in the perilymph. He observed an eddy current which, now bearing his name, has dominated speculation on mechanisms of the cochlea ever since. Recently, Inselberg (Inselberg et al., 1975) has suggested that the eddy of von Békésy is artifactual. On the basis of our macroscope results, it would seem that the elastic macron—which was invisible to von Békésy—is more likely than the simmering macron he saw to be the mechanism of hearing. This question is the subject of current research.

We shall now consider the brain from the macron point of view. As a physical object, it is apparently a bioplasmic spacework with hierarchical structure. Its very physical structure suggests an elastic macron with clearly defined nodal surfaces. Its various segments support compound physical (elastic), chemical, and electrical macrons—which are coupled through the dendritic surface. So far, there has been no discussion of a functional role for the elastic vibrations of the brain body. But from the macrodynamic point of view, the elastic behavior is coupled to the electrochemical state through known plasma mechanisms, and probably also through liquid crystal (flexoelectric) mechanisms as well, so the possibility of a functional role cannot be ignored. In any case, what can be said at this stage is just a conjecture: *a thought is a macron of the brain bioplasma.* This suggests a physical mechanism for a holistic approach to brain function, and is certainly at odds with the connectionist theory of the neural network. This conjecture leads easily to more precise conjectures for specific brain functions. For example, the transfer from short-term to long-term memory might be explained as follows: A short-term memory is a brain–body macron, metabolized (or

ISBN 0-201-03438-7/0-201-03439-5pbk.

driven) by the neural network. This macron maintains a spatial pattern of various biochemical and ionic particles, as in the Howard–Kopell phenomenon. As this pattern is maintained through repeated neural activation of the macron (thought), some molecules within this pattern become attached to membranes and thus immobilized. This physical realization of the engram (macron) pattern is the long-term memory. Recall is effected by a macron resonance phenomenon; and so forth.

The juxtaposition of these two examples of macrodynamic processes, hearing and thinking, suggests that the whole information-processing chain can be interpreted as a flow of macrons extending, through coupling, across different media. This idea has been carried to extremes by Thom (1973), in his psycholinguistic theory.

The morphodynamic conjecture for brain function is not about to be established by any current research program. Yet there is some work on electrical macrons in the dendritic surface—that is, spatial patterns of EEG potentials. Various results (F. Abraham, 1973; Adey, 1974; Brazier, 1969; Freeman, 1975) suggest that brain macrons have functional roles. As a last laugh, we propose that the classical salty jelly experiment of Kennedy (1961), supposedly ridiculous, has serious implications.

Noology.

Probably the macrodynamic brain theory has eliminated all but the most credulous readers. If there are any survivors, we may as well dispose of them now by discussing the macrodynamics of consciousness. Actually, this is not impossible, as there exists a (quantum) mechanical theory of consciousness, thanks to Walker (1970), which admits of a macrodynamic formulation. However, let us ignore the question of mechanism. Suppose a human being can be identified with a conscious unit, a particle in the noosphere. Suppose, furthermore, that these macrodynamic units are coupled by communication, a macron resonance phenomenon, as described in the brain speculation. Then, the noosphere may be described as a complex system of macrons. Actually, this idea can be formalized mathematically, so that an archetype in the collective unconscious becomes a stable elastic vibratory state of the noosphere (the Big Salty Jelly in the Sky). This provides macrodynamic mechanisms for astrology, telepathy, clairvoyance, synchronicity, and so forth, as I have proposed in "Psychotronic Vibrations" (R. Abraham, 1973). The existence of a new force, the psychotronic field, is a separate question.

ISBN 0-201-03438-7/0-201-03439-5pbk.

3. CONCLUSION

In this introduction to macrodynamics and its applications, there is admittedly an inordinate amount of speculation. I regret this sincerely, but as many have discovered, speculation is much faster than experimental work. I must therefore single out especially the hardware projects, described in the section on techniques of macroscopy, as terra firma in this ocean of dreams. And I confess that my goal in these hardware projects, in addition to my own curiosity, is a political one: to stimulate a wave-conscious, morphological orientation in scientific research.

REFERENCES

Abraham, F. (1973). "Spectrum and Discriminant Analyses Reveal Remote rather than Local Sources for Hypothalamic EEG: Could Waves Affect Unit Activity?," *Brain Research*, **49**, 349–366.

Abraham, R. (1972). "Introduction to Morphology," *Publ. du Dept. de Mathématique*, **9**, 38–174. Lyon: Université de Lyon.

Abraham, R. (1973). "Psychotronic Vibrations," *Proc. First Intern. Congr. Psychotronics*, Prague.

Adey, W. (1974). "The Influence of Impressed Electrical Fields at EEG Frequencies on Brain and Behavior," preprint, Math. Inst.,

Anthony, P. (1969). *Macroscope*. New York: Avon.

Békésy, G. von (1960). *Experiments in Hearing*. New York: McGraw-Hill.

Brazier, M., and Walter, D. (eds.). (1969). *Advances in EEG Analysis*. New York: American Elsevier.

Freeman, W. (1975). *Mass Action in the Nervous System*. New York: Academic Press.

Haeckel, E. (1974). *Art Forms in Nature*. New York: Dover.

Hirsch, M., and Smale, S. (1974). *Differential Equations, Dynamical Systems, and Linear Algebra*. New York: Academic Press.

Howard, L., and Kopell, N. (1976). "Pattern Formation in the Belousov Reaction," *Proc. AMS-SIAM Symp.*

Inselberg, A., Chadwick, R., and Johnson, K. (1975). "Mathematical Model of the Cochlea." *SIAM J. Appl. Math.*

Isnard, C. A., and Zeeman, E. C. (1975). "Some Models from Catastrophe Theory in the Social Sciences," in *The Use of Models in the Social Sciences* (Lyndhurst Collins, ed.). Boulder, Colorado: Westview Press.

Jenny, H. (1967). *Kymatik*, Bd. 1. Basel: Basileus.

ISBN 0-201-03438-7/0-201-03439-5pbk.

Jenny, H. (1972). *Kymatik*, Bd. 2. Basel: Basileus.

Kennedy, J. (1961). "A Possible Artifact in the EEG," *Psychol. Rev.*, **66**, 347-353. See also Oswald (1961).

Marsden, J., and McCracken, M. (1976). *The Hopf Bifurcation and its Applications.* New York and Berlin: Springer.

Oswald, I. (1961). "On the Origin of the α Rhythm," *Psychol. Rev.*, **68**, 360-362.

Ruelle, D., and Takens, F. (1971). "On the Nature of Turbulence," *Communs. Math. Phys.*, **20**, 177-192; **23**, 343-344.

Settles, G. (1971). "The Amateur Scientist," *Scientific Amer.*, May 1971. See also Stong (1974).

Stong, C. (1974). "The Amateur Scientist," *Scientific Amer.*, Aug. 1974.

Thom, R. (1973). *Stabilité Structurelle et Morphogenèse* (3d printing, with corrections). Reading, Mass.: Benjamin Advanced Book Program. See also *Structural Stability and Morphogenesis* (D. H. Fowler, transl.; rev. and updated by the author). (1975). Reading, Mass.: Benjamin Advanced Book Program.

Thompson, D'Arcy, W. (1945). *On Growth and Form*, 2 vols. Oxford: Cambridge Univ. Press. Second ed., 1963; abridged ed., 1961.

Turing, A. (1952). "A Chemical Basis for Biological Morphogenesis," *Phil. Trans. Roy. Soc. (London), Ser. B.*, **237**, 37.

Waddington, C. H. (ed.). (1968-1972). *Towards a Theoretical Biology.* 4 vols. Edinburgh: Edinburgh Univ. Press; Chicago: Aldine (Vols. 1, 2); New York: Halsted Press (Vol. 3).

Walker, E. H. (1970). "The Nature of Consciousness," *Math. Biosci.*, **7**, 131-178.

Zeeman, C. (1971). "The Geometry of Catastrophe," *Times Lit. Suppl. (London)*, **1556**, issue of 10 Dec. 1971.

ISBN 0-201-03438-7/0-201-03439-5pbk.

7

Simulation of Self-Renewing Systems

Milan Zeleny and Norbert A. Pierre

1. AUTOPOIETIC ORGANIZATION

Autopoietic systems are self-renewing, self-repairing, and unity-maintaining autonomous organizations of components capable of interactive linkages. Examples are cells, organs, organisms, and groups of organisms. *Autopoiesis*, or self-creation, characterizes all living organisms and their organizations, ranging from the macromolecular, unicellular and multicellular organisms to differentiated, self-perpetuating animal and human groupings.

Autopoietic organization can be defined as a network of interrelated component-producing *processes* such that the components, through their interaction, generate recursively the *same* network of processes which produced them and thus realize the network of processes as an identifiable unity in the space in which the components exist. The product of an autopoietic system is necessarily always the system itself, its organization being continuously realized under permanent turnover of matter and energy.

Allopoietic organization, in contrast, can be defined as a network of interrelated component-producing processes such that it does not produce the components and processes which realize it as a unity. The product of an allopoietic system is different from the system itself. Thus, the actual realization of such systems is determined by processes which do not enter into their organization. They are nonautonomous, since their realization and longevity as unities are not related to their operation. Among the examples here figure spatially determined structures like crystals, formal hierarchies, and concentration camps. The process of production of the components that realize the allopoietic organization as a unity does not enter into its definition as a unity.

The simplest autopoietic organization is a *cell* of a biological organism, not because there could not be other autopoietic or allopoietic systems within the cell, but because it clearly displays the *minimal* or-

Erich Jantsch and Conrad H. Waddington (eds.), *Evolution and Consciousness: Human Systems in Transition.*

ISBN 0-201-03438-7/0-201-03439-5pbk.

ganization, for *any* set of components, necessary for autopoiesis. There is a *catalytic nucleus* capable of interactions with environmental substrate so that the membrane-forming components are produced. A *membrane* then defines and separates this network of interactions from its environment and an autonomous unity is thus realized. A cell is a continuous and recursive production of components which, through a membrane, define the cell itself.

Although there are myriad distinct subcellular structures within cells—atoms, molecules, macromolecular polymers, mitochondria, chloroplasts, and so forth—the properties of the components do not determine the cell's properties as an autopoietic system. The properties of a cell are the properties of relations and interactions produced by, and producing, its components.

2. AUTOPOIETIC MODELING

Autopoietic modeling has generically evolved from cellular automata theory (see e.g., von Neumann, 1966; Codd, 1968; and Gardner, 1971). Recently, Wainwright (1974) has reviewed the Game of Life, an interesting tessellation model governed by only two simple rules.

In 1974, three Chilean scientists (Varela et al., 1974) published a seminal article describing for the first time a truly autopoietic model, representing a new direction as well as a new hope in contemporary biology. The article is equally important to simulation modeling of social systems, organizational theories, and management sciences.

In accordance with the basic organization of a cell, the simplest model of autopoietic organization must consist of an environment of substrate, catalysts capable of producing more complex elements (i.e., links), which are capable, in turn, of their own bonding, concatenating into a membrane around the catalyst. Including holes, there are thus five different states possible in each position of the grid, because each position can be occupied by one of the five basic components only, which are assigned the following graphical symbols:

Symbol	Component
	hole
○	substrate
▯	link
-▮ -⊟-	bonded link (singly or fully)
★	catalyst

In the cell model of Varela et al. (1974), three basic transformations are accounted for:

ISBN 0-201-03438-7/0-201-03439-5pbk.

1. *Composition:* 2○ + ★ ⟶ □ + ★
   A catalyst and two units of substrate produce a link, while the catalyst is left unchanged. A hole is the byproduct of this operation.
2. *Disintegration:* □ ⟶ 2○; -□ ⟶ 2○; -□- ⟶ 2○
   A link, free or bonded, disintegrates into two units of substrate, filling available holes.
3. *Bonding:* □-□- ··· -□ + □ ⟶ □-□- ··· -□-□
   A free link can be bonded with a chain of bonded links; two chains of bonded links can be bonded into one; two free links can be bonded to start a chain formation.

The actual rules guiding the movement of all components and specifying the conditions for the three basic transformations are quite involved and are formally treated elsewhere (Zeleny and Pierre, 1975). However, it should be mentioned here that, besides the basic system-building components just enumerated the simple model includes two other main constituents:

(a) A simulation of chance, based on a random number generator functioning in conjunction with a special set of rules, including a random walk over two- or three-dimensional space in the tessellation grid (i.e., randomly determining the direction of movement or, alternatively, the next configuration to be established by movements).
(b) Recursive transformation and interaction rules or "algebra," allowing spontaneous encounters and linking of elements as well as subsequent disintegration of the created linkages.

Each link is allowed to have only two bonds: it can be either singly bonded, -□ , or fully bonded, -□- . Additional bonds are not allowed, to avoid branching of chains. Such unbranched chains can ultimately form a membrane around the catalyst, creating an enclosure which is not penetrable by either ★ or □ . These components are thus effectively "trapped" and forced to function for the benefit of the autopoietic unity. Substrates ○ can pass freely through the membrane to facilitate production of □ by the catalyst. Any disintegrated links, causing ruptures in the membrane, are thus effectively repaired by this continuous production. The unity of the system is dynamically reestablished.

Multiple catalysts can function in the same tessellation grid of substrate. As the catalysts move closer together or as their number increases, a dynamic balance among interacting membranous unities—a *multicellular autopoietic system*—can be established. Similarly, we can control available amounts of substrate inflows by simply contracting or expanding the corresponding tessellation grid. The rates of membrane formation and disintegration can also be regulated by a simple adjust-

ISBN 0-201-03438-7/0-201-03439-5pbk.

ment of particular rules and parameters. The system can be "forced" to disintegrate totally or to freeze into stable allopoietic structure. Systems incapable of even forming a membrane can be induced into their futile existence. Systems whose membranes never rupture; systems with spacious, large, or narrow membranes; substrate-seeking "amoeba-like" cells, floating through the space; and hundreds of other varieties can be observed by simply adjusting a few rules.

Extraordinarly intricate phenomena can be observed to be *spontaneously generated* by this simple set of rules. In our first simulation studies with a model, coded in APL for the IBM-360 system at Columbia University (see Figure 7.1), the autopoietic "birth–death" process ultimately attains a dynamical steady state, characterized by a self-renewing "membrane." A cathode-ray tube graphical image then allows for continuous observation and analysis of its "life." Inducing a proper deviation from such a steady state can cause disintegration of the "membrane" and ceasing of all vital functions. Similarly, "explosive" conditions, leading to "cancerous" growth, can be induced.

In Figure 7.1, it may be observed that after reaching dynamic stability (somewhere around TIME 25, i.e., just before the first closure of the membrane), the autopoietic cell keeps renewing itself through a series of oscillations between rupture and closure. Its very existence as an autopoietic system is based on this rhythmical opening and closing. We observe that the underlying rules have created a "natural rhythm" of the open system (see, e.g., Klapp, 1975). All living systems, including societies, function through a complex of more or less intricate biorhythms. We might preferably talk of *pulsating systems*, since neither permanently closed nor permanently open systems are autopoietic; they are not "alive."

3. THE AIM: PARAPHRASING DYNAMICAL ASPECTS OF A SYSTEM

The conventional method of reduction approaches complex systems by breaking them up into smaller components. If these are still too complex, they are broken up into even smaller components, until the resulting components are so tiny that at least one of them can be understood. There are roughly 10^{20} molecules in a single living cell. In the cell's nucleus alone, more than one hundred distinct chemical reactions have been identified; yet the properties of components in isolation add little, if anything, to our understanding of the workings of a cell.

Cancer is a problem of this extraordinarily complex universe, the cell. It is due to the disruption or alteration of the autopoietic network

ISBN 0-201-03438-7/0-201-03439-5pbk.

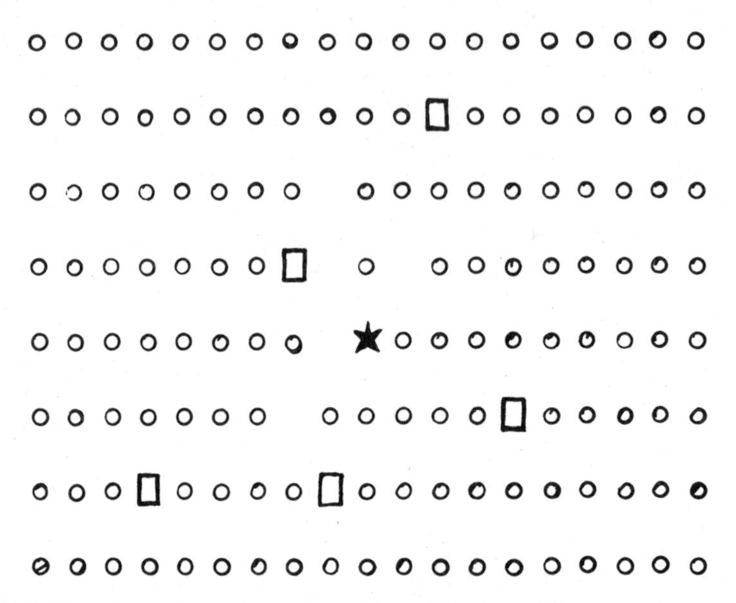

FIGURE 7.1. Simulation of an autopoietic cell system. (The symbols are explained in the text.) (a) TIME 5. The process is initiated by putting a catalyst in the environment of substrate; first free links are produced.

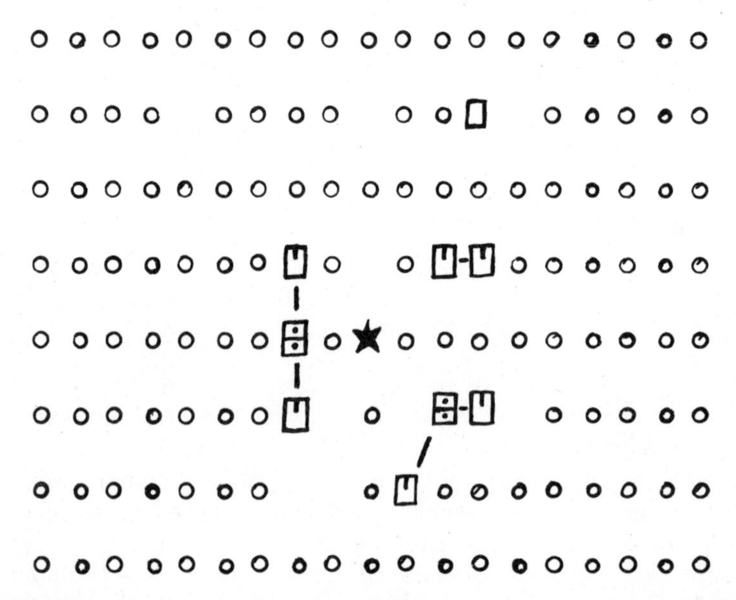

FIGURE 7.1b. TIME 10. Catalytic production continues and some initial bonding occurs. Observe that all matter is conserved: the number of holes is always equal to the number of all links (free and bonded).

ISBN 0-201-03438-7/0-201-03439-5pbk.

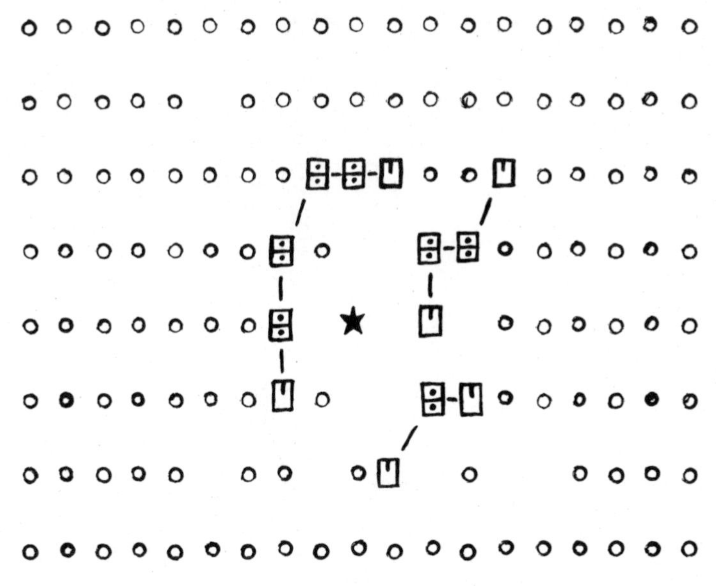

FIGURE 7.1c. TIME 15. Larger chains of bonded links are being formed. These concatenations already tend to be positioned around the catalyst.

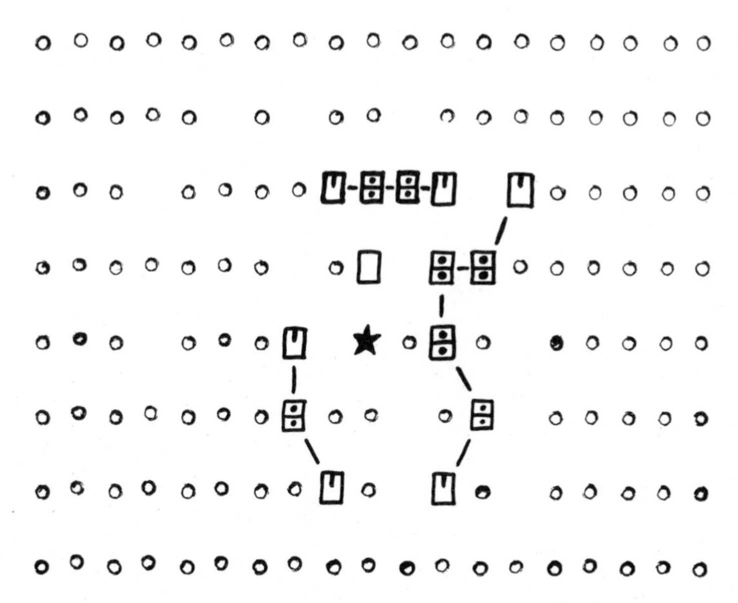

FIGURE 7.1d. TIME 20. Even though some links disintegrate in the process, we can observe an overall increase in the amount of organized components.

ISBN 0-201-03438-7/0-201-03439-5pbk.

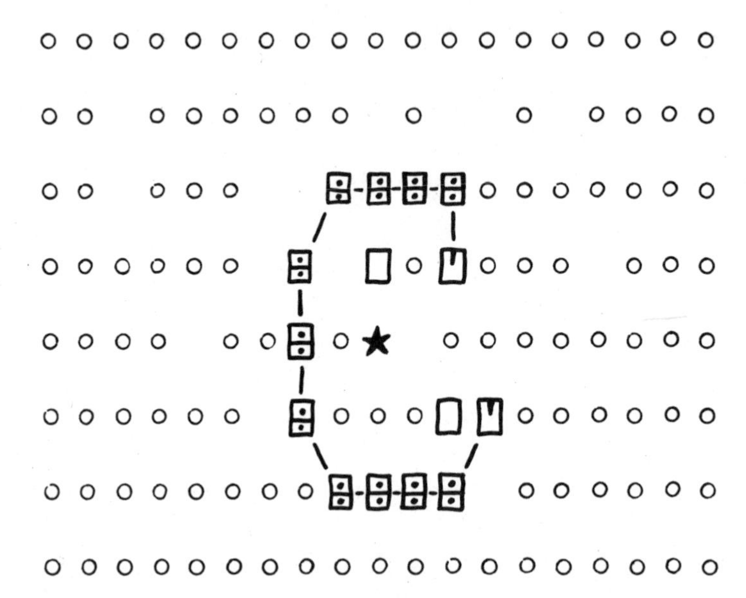

FIGURE 7.1e. TIME 25. The ratio of links to substrate, which is now almost one eighth, indicates that the "life-sustaining conditions" have been approached. Next we look at the succeeding iteration (rather than screening only every fifth one as before).

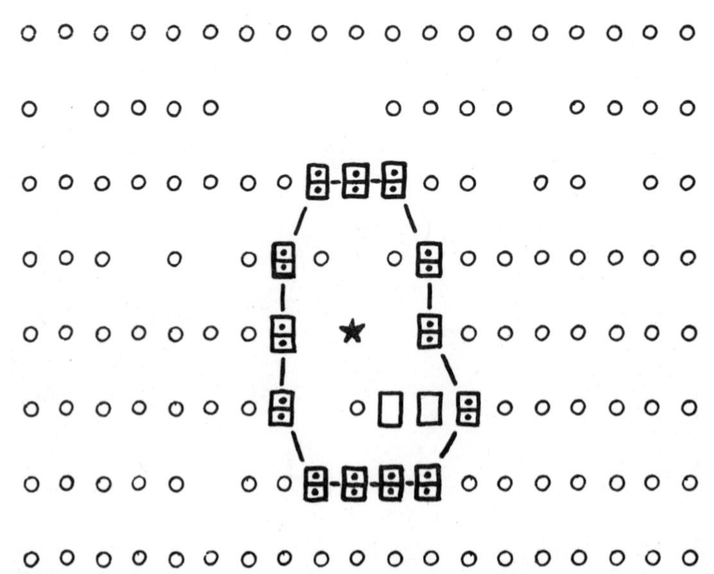

FIGURE 7.1f. TIME 26. A membrane has enclosed itself around the catalyst and a clearly defined unity of a certain size and shape can be observed. We now look at every second iteration.

ISBN 0-201-03438-7/0-201-03439-5pbk.

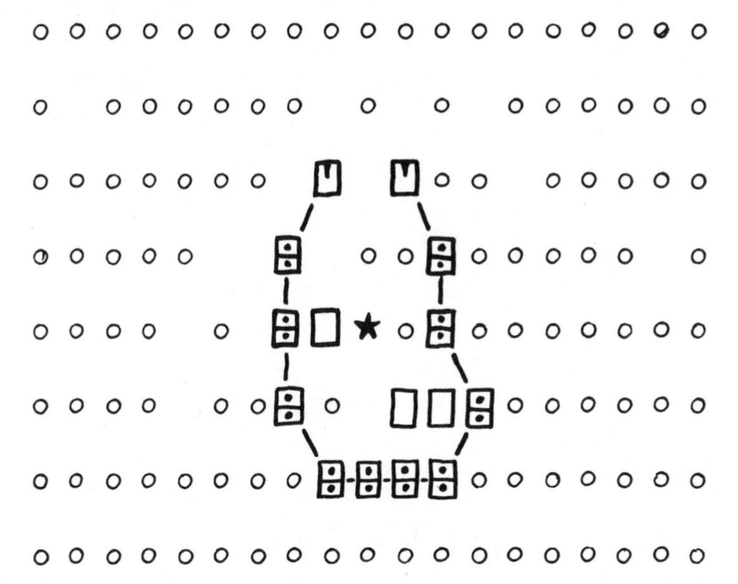

FIGURE 7.1g. TIME 28. The membrane has been ruptured. Although the essential forms and functions are still being perpetuated, the unity must repair itself in order to be truly autopoietic.

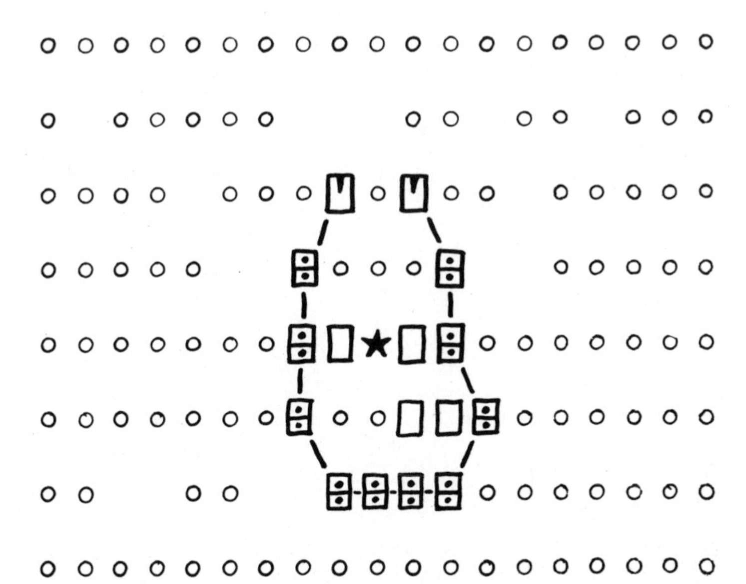

FIGURE 7.1h. TIME 30. Observe the increased number of links and substrate concentrating inside the cell. The chances for an early reinstatement of the whole membrane are good.

ISBN 0-201-03438-7/0-201-03439-5pbk.

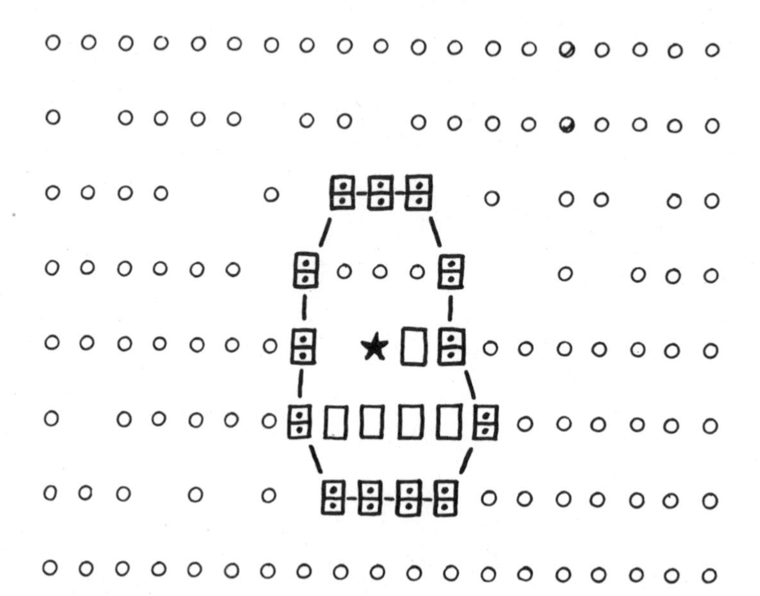

FIGURE 7.1i. TIME 32. The membrane has been repaired and the autopoietic unity restored.

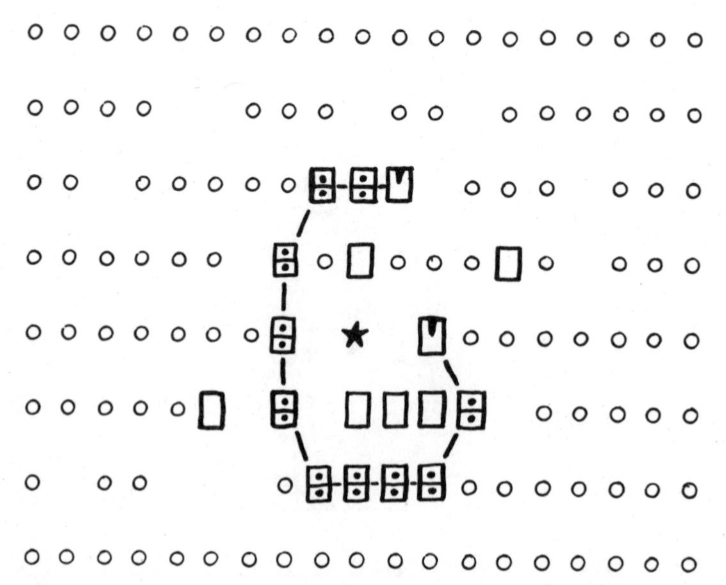

FIGURE 7.1j. TIME 34. We can observe another rupture of the membrane.

ISBN 0-201-03438-7/0-201-03439-5pbk.

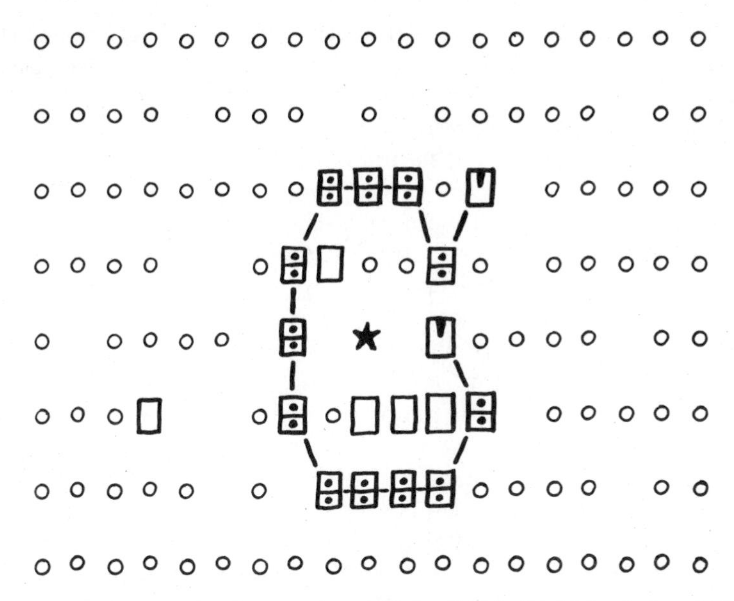

FIGURE 7.1k. TIME 36. The number of links (19) is the highest now and early repair is imminent. The autopoietic unity is going to "survive" until we turn the computer off. Next we look fourteen iterations ahead.

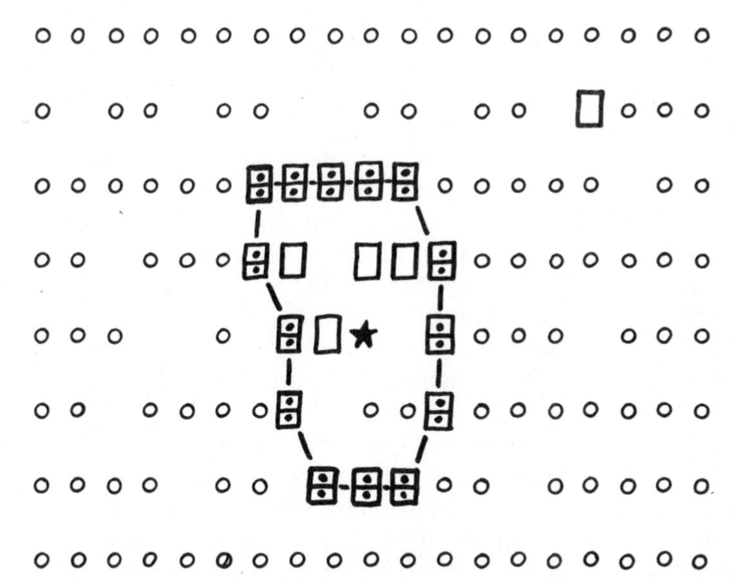

FIGURE 7.1l. TIME 50. The unity is still clearly defined and the whole autopoietic process exhibits global stability of a series of ruptures/repairs. Note the changed contour of the membrane.

ISBN 0-201-03438-7/0-201-03439-5pbk.

of interactions. Yet reductionists hope to gain the crucial understanding of the functions (or dysfunctions) involved by studying its tiniest components: molecules of ribonucleic acid, viruses, chemical reactions at the molecular level, and so forth.

A cell, as an autopoietic system, cannot be understood by studying the properties of the components. Its properties as a whole are determined by the properties of the *interactions* between the components, that is, by its dynamical organization. To try to ascribe a determinant value to any component, or to any of its properties, be it DNA, RNA, or a virus, is a scientific artifice.

In this situation, we face a formidable dilemma: We cannot afford to lose sight of the full complexity of the cell, and on the other hand, it becomes increasingly difficult to cope with this complexity in our inquiry. The number of chemical reactions that must be going on simultaneously in a cell at a given instant of time simply staggers the imagination. If these complex processes could be translated into thousands of corresponding differential equations, a mathematical model of the cell, we would have a tool which could simulate chemical reactions and test suspect aberrations as well as hypothesized control actions. However, even the most sophisticated solution procedures in conjunction with the most powerful computer systems will allow solutions of, at most, only fifty to sixty differential equations—a tiny fraction of the interactions within a cell, indeed.

Foerster (1971) has proposed another avenue by which to bypass this dilemma: "The idea is to abandon the strategy of reformulating the problem into terms that smack of mathematical rigor but lack the contextual richness originally perceived, and to develop the algorithms that transform the descriptions of certain aspects of a system into paraphrases that uncover new semantic relations pertaining to the system as a whole." The rules, transformations, transitions and interactions observed in Conway's Game of Life and in the autopoietic model outlined in the preceding section—they are the paraphrases! The set of computer-based rules governing the organization in its autopoietic state has an enormous potential. It is "bioalgebra," a formal tool for studying living and bioadaptive systems, an autopoietic mathematics that "lives" as does the system it describes.

Instead of studying details of separate chemical reactions, we must concentrate on the interaction of autopoietic components: the set of rules, or program; process generators, transducers monitoring the output of the process generators; autopoietic regulators; repressors; and so forth. Autopoietic models allow us to study their interactions in a holistic and organic way. Cybernetic models are bound to fail here because they attempt to define communication channels, or "wired circuits," of

ISBN 0-201-03438-7/0-201-03439-5pbk.

pathways linking these components—an impossible task even if the cell were an electronic or cybernetic system.

Through the autopoietic model we could plot the functions and amounts of main autopoetic components and draw "profiles" of cells under different environmental and organizational conditions. These profiles should allow us to detect the main differences between normal and cancerous cells in terms of their overall interactive patterns. We could then test the balance-restoring actions on the computer many times before trying them on patients. We could treat cancer without being concerned about the chemical-molecular properties of the balancing and restorative agents.

To give an idea of the complexity involved in setting up simple autopoietic models, we might mention that we have found that the autopoietic model of Varela et al. (1974) can be programmed and operated by graduate students in less than a month.

4. APPLICATION TO HUMAN ORGANIZATION

The design of allopoietic institutional structures should not be confused with the study of autopoietic organizations. To specify spatial relations between components, to create a frozen image, a pyramidal hierarchy, has probably little in common with growing an autopoietic organization.

The task of management is to stimulate the *growth of a network of decision processes*, systems, programs, and rules, that is to say, an organization which may be considered effective in attaining institutional objectives. Since one of these objectives is the continuous self-renewal of the autonomous dynamic unity of the organization (i.e., an autopoietic operation), the network of decision processes must produce components capable of recursively generating the same network through their interaction. In this sense, *a manager is the catalyst* rather than the designer of an organization.

Mechanistic concepts of organization and of the design or redisign of its "mechanisms" are the concern of the modern *cybernation* approach, described by Beer (1975, p. 107) as follows: "Thus, we shall gradually be able to devise a more complicated model, redolent with feedback loops, which is of practical value. In doing all this, we pass from the notion of a straightforward feedback mechanism to the notion of multiple loop systems." What we are left with after such strenuous modeling could be just some *allopoietic debris.* No matter how many loops are designed, no matter how complex they are, if they are not

ISBN 0-201-03438-7/0-201-03439-5pbk.

self-productions of the organization itself, there is no autopoietic, autonomous, and dynamical system in existence.

In order to have autopoietic systems, the components must exhibit a multitude of interactions through their decision-making capabilities. Human systems differ from all other systems in the enhanced ability of their components to make decisions, to choose among or to create alternatives according to their own objectives. Yet the notions of *goals* and *decision-making* are missing from both cybernetics and theories of organizational behavior. The latter are designed to control and to predict the *behavior* of organizations. They attempt to elicit certain behavioral patterns by imposing the rules or designing a structure from without.

Hayek (1975) characterizes the order of social events as such that, though it is the result of human action, it has not been created by men deliberately arranging the elements in a preconceived pattern. He also talks about ordering forces, spontaneous orders, and "the rules." If we understand the forces that determine such an order, we can use them by creating the conditions under which such an order will form itself. The autopoietic approach has the advantage that it can be used to procure orders that are far more complex than any order we can produce by arranging the individual components in their appropriate places—a method for systems analysis. Hayek (1975, p. 11) actually preconceives a *social autopoiesis* when he writes: "Though the conduct of the individuals which produces the social order is guided in part by deliberately enforced rules, the order is still a spontaneous order, corresponding to an organism rather than to an organization. It does not rest on the activities being fitted together according to a preconceived plan, but on their being adjusted to one another through the confinement of the action of each by certain general rules."

The attained degree of complexity of the structure of modern society exceeds by far that which would be possible to achieve by deliberate organization. It is a paradox of a futurist contending that we must deliberately plan modern society because it has become so complex. Rather it turns out that we can preserve an order of such complexity only if we control it not by the method of "planning" (i.e., by direct orders), but on the contrary, by aiming at the formation of a spontaneous order based on general rules—a strategic *scenario* planning at best.

Modern computer and simulation methodology are powerful tools with which to grow and observe real social organizations in laboratory settings. To use these great advances effectively, however, we have to treat values, objectives, decisions, norms, and other precepts of human conduct as essential attributes of human systems and their manage-

ISBN 0-201-03438-7/0-201-03439-5pbk.

ment. By reducing human conduct to a succession of overt physical behaviors, modern behaviorists and behavioral science have prevented themselves from coping successfully with so important a social phenomenon as autopoietic organization. They are bound to endless batch-processing of quantities of primitive data.

The idea of *computer simulation of organizations* can produce only trivial results in dealing with allopoietic behavioral systems. Such "behavioral simulation" can only predict or control the behavior of organisms whose actions rigidly follow rules imposed by the designer from without. In contrast, managers as catalysts induce the components to make their own decisions, conduct their own analyses, select their own criteria. A unique autopoietic organization, a network of values, norms, and precepts, is self-created, self-maintained, and self-grown.

Such a growth process of an autopoietic unity evolves its own rules of change. These rules, in turn, determine what kinds of development can occur. The totality of such an autopoietic network of rules constitutes a theory of organizations. Social change results from free choice by independent decision-makers; it is a teleological advance and need not be just the shaped outcome of external environmental pressures. The pioneers of social simulation, Beatrice and Sidney Rome (1971, p. 2), state,

> When people form and become members of an organization, they agree to conduct themselves according to rules or prescripts that *they* establish and maintain. These covenants stipulate what members must, may, can, ought, and should do and not do, according to their roles in the organization, and these enable individuals to act as agents for the corporate whole. As organizations develop, their covenants become successively transformed. Inasmuch as these processes have formal order, a science of social development is possible, that depends not on past regularities, but on future opportunities.

But human systems are not only contrivances for processing information and making decisions. Humans *live* their lives through human systems, shape them through their *individual* aspirations, goals, norms, values, and actions, creating a set of *systems* aspirations, goals, norms, values, and actions, which could be quite different from and independent of the individual ones. Humans are, in turn, continuously being shaped by these self-organized entities, their spatial arrangement evolving through a succession of orderly but temporary structures. Human purposeful action and the autopoietic interaction with their emerging organization are *irreducible to behavior*, as has been so forcefully stated by Jantsch (1975).

ISBN 0-201-03438-7/0-201-03439-5pbk.

5. TOWARD A HUMAN SYSTEMS MANAGEMENT

Human systems management requires a new mode of inquiry into complex and dynamical human systems. Its contours emerge in the context of the following set of observations:

1. Complex and dynamical human systems are to be *managed* rather than analyzed or designed. Human systems management is not systems analysis or design.
2. Human systems management is a process of *catalytic reinforcement* of a dynamic self-organization and bonding of human components. It does not design a managerial hierarchy of command or control.
3. Components of human systems are *humans.* As such, they differ significantly from other components, mechanistic or biological, in their ability to anticipate the future, to formulate their objectives, to plan for their attainment, and to make decisions. These properties are sufficient to make human systems quite distinct from all other systems.
4. The integral complexity of human systems can be lost in the process of its simplifying reinterpretation by the rigor of mathematical mechanics. Human systems can be described and studied through a relatively simple set of linguistic, fuzzy, and semantic rules, governing the self-creation of its complex organization. Human systems management is not operations research, econometrics, or applied mathematics.
5. Interactions between components are not those of electronic circuitry, communication channels, or feedback loop mechanisms. Rather, they are organic and dynamical manifestations of organizational autopoiesis. Human systems management is not cybernetics or the information theory of communications.
6. Dynamical order of human systems organization is maintained through a continuous renewal of certain *nonequilibrium conditions.* Both nonequilibrium and instability are essential for self-organization of higher complexity. Human systems management is not a theory of general equilibrium.
7. The concepts of optimization and optimal control are not meaningful in a general theory of human systems. Human aspirations and objectives are dynamical, multiple, and in conflict, as are those of human organizations. This conflict is the very source of their creative evolutionary unfolding. Human systems management is not optimal control theory or theory of conflict resolution.
8. The inquiry into human systems is *transdisciplinary* by necessity. Human systems encompass the whole hierarchy of natural systems: physical, biological, social, and spiritual. Human systems management is not interdisciplinary or multidisciplinary; it does not attempt to unify scientific disciplines, but transcends them.

In conclusion, we believe that autopoietic modeling, as outlined earlier in this chapter, carries great potential for making a significant contribution to the development of human systems management in this spirit.

ISBN 0-201-03438-7/0-201-03439-5pbk.

REFERENCES

Beer, Stafford (1975). *Platform for Change.* New York: Wiley.

Codd, E. F. (1968). *Cellular Automata.* New York: Academic Press.

Foerster, Heinz von (1971). "Computing in the Semantic Domain," *Ann. N. Y. Acad. Sci.*, **148**, 239–241.

Gardner, M. (1971). "On Cellular Automata, Self-Reproduction, the Garden of Eden, and the Game of 'Life'," *Scientific Amer.*, **224**, No. 2, p. 112.

Hayek, Friedrich A. von (1975). "Kinds of Order in Society," *Studies in Social Theory, No. 5.* Menlo Park, Calif.: Institute for Humane Studies.

Jantsch, Erich (1975). *Design for Evolution: Self-Organization and Planning in the Life of Human Systems.* New York: Braziller.

Klapp, Orrin E. (1975). "Opening and Closing in Open Systems," *Behavioral Sci.*, **20**, 251–257.

Neumann, John von (1966), in *The Theory of Self-Reproducing Automata*, (A. W. Burks, ed.) Urbana, Ill.: Univ. of Illinois Press.

Rome, Beatrice K., and Rome, Sidney C. (1971). *Organizational Growth through Decisionmaking.* New York: American Elsevier

Varela, F. G., Maturana, H. R., and Uribe, R. (1974). "Autopoiesis: The Organization of Living Systems, Its Characterization and a Model," *Bio-Systems*, **5**, 187–196.

Wainwright, R. T. (1974). "Life is Universal!," *Proc. 1974 Winter Computer Simulation Conference (New York).* La Jolla, Calif.: Simulations Council.

Zeleny, Milan, and Pierre, Norbert A. (1975). "Simulation Models of Autopoietic Systems," *Proc. 1975 Summer Computer Simulation Conference (San Francisco).* La Jolla, Calif.: Simulations Council.

ISBN 0-201-03438-7/0-201-03439-5pbk.

Part III

Aspects of Socio-cultural Evolution

In consequence of the autonomy of the physical phenomena there cannot be only one *approach to the mystery of being—there must be at least two: namely, the physical happening on the one hand, and the psychic reflection on the other; but it is hardly possible to decide what is reflecting what!*

Carl Gustav Jung

Self-transcendence in the human world expresses itself through the complementarity of purposeful action and emergence. The history of mankind may be understood as an ongoing process of aligning conscious design with evolutionary emergence, a process which is made possible by an increasing capability of dealing flexibly with the unexpected, with uncertainty, variability, and change. In our day, this process has reached a stage where purposeful cultural design becomes the new great task. But rather than aiming at specific cultural structures, this new level of human design will focus on interactive processes, on cultural pluralism, symbiosis, and a *dynamic balance* in terms of quasi-continuous qualitative change transcending the historical succession of unique cultural regimes. Such a *metaregime* will mark the beginnings of man's flexible correlation with a new level of environment: mankind-at-large.

E. J.

ISBN 0-201-03438-7/0-201-03439-5pbk.

Erich Jantsch and Conrad H. Waddington (eds.), *Evolution and Consciousness: Human Systems in Transition.*

8

Process and Structure in Sociocultural Systems

Alastair M. Taylor

1. INTRODUCTION: THE NEED FOR A NEW PARADIGM

Crystals abound in planes of symmetry but are inert, whereas organisms capable of the greatest self-motion possess only bilateral symmetry. That an inverse ratio seems to exist between symmetry and mobility is in keeping with Dagobert Frey's observation' "Symmetry signifies rest and binding, asymmetry motion and loosening, the one order and law, the other arbitrariness and accident, the one formal rigidity and constraint, the other life, play and freedom" (Frey, 1956).

Such an observation is apposite because there has been a strong tendency in the literature of sociological theory and the social sciences in general to emphasize—both conceptually and normatively—the role of symmetry in societal structures and equilibrium in societal processes. We might cite the functionalist approach (as exemplified by Talcott Parsons), which conceptualizes societies as systems of interrelated parts with causation at once multiple and reciprocal. Although integration is never complete, social systems are viewed as fundamentally in a state of "dynamic equilibrium," making adaptive responses either to exogenous change or to endogenous change in one of its parts. Change occurs generally in a gradual, adjustive fashion within a society whose chief source of integration is a system of values to which most of its members adhere. The tensions and conflicts which exist are regarded as aberrations from the "dominant pattern"; however, such "deviance" tends to be resolved or institutionalized in the long run (van den Berghe, 1963).

From the foregoing, it can be seen that this brand of functionalism adheres to a concept of equilibration that emphasizes continuity, gradualness, and uniformity in matters of change—and indeed leaves Parsons open to the charge that "his is a conservative doctrine that sees social change as deviant to social order, and as a phenomenon which is possible only when the 'control mechanisms of the social system' break down" (Horowitz, 1962). Meanwhile, although the dynamic equilib-

ISBN 0-201-03438-7/0-201-03439-5pbk.

Erich Jantsch and Conrad H. Waddington (eds.), *Evolution and Consciousness: Human Systems in Transition.*

rium model has proved a useful instrument in dealing with two types of change—growth in complexity through differentiation, and adjustment to extrasystemic changes (e.g., problems of acculturation)—we agree with van den Berghe (1963; see also Demerath and Peterson, 1967) that it cannot satisfactorily account for the "irreducible facts" that:

1. reaction to extrasystemic change is not always adjustive;
2. social systems can, for long periods, undergo a vicious circle of ever-deepening malintegration;
3. change can be revolutionary, that is, both sudden and profound; and
4. the social structure itself generates change through internal conflicts and contradictions.

As Jantsch has pointed out in Chapter 3, social theories have been traditionally geared to structure, not process, and to ideals of equilibrium and structural stability. Hence the emphasis upon steady state (negative feedback) and, given our contemporary ecological problems, "limits to growth." However, in our time these traditional equilibrium models fail to represent adequately the dynamics of contemporary human society, which is in a state of seemingly progressive dysfunctionalism in a planetary environment that in turn is undergoing "sudden and profound" ("revolutionary") change. Bringing to our section of this volume concepts raised by Holling (Chapter 4) and Prigogine (Chapter 5), we might ask the following questions:

1. What is the relationship of societal stability to structural complexity?
2. What is its relationship to process, that is, to resilience or persistence?
3. How do we account for societal quantization (discontinuities) within an overall evolutionary continuum?
4. Are we presently in a societal-environmental transformation that needs to be conceptualized in terms of nonequilibrium dynamics?

2. CONSTRUCTING A SOCIOCULTURAL SYSTEMS MODEL

A *system* comprises a constitutive complex whose characteristics depend on the specific relations of its parts. It also exists in an enveloping environment with which it interacts. A system equilibrates with its environment by means of *feedback*—with the negative type correcting deviations in the existing system–environment relation, whereas positive feedback amplifies deviations from the existing "norm" (as in the case of growth or, again, decline conditions in a system). In our model, we shall be examining the interaction of these feedback proc-

ISBN 0-201-03438-7/0-201-03439-5pbk.

esses in terms of a sociocultural system's capacity to (1) exercise some kind of environmental control, and (2) persist in time. We must also be able to account for systemic and environmental transformations that are associated with different levels of organizational complexity among human societies.

Negative and positive feedback processes can be correlated with systemic continuity/discontinuity. Continuity results when deviation-correcting mechanisms operate so as to ensure that structure and function remain viable within the given parameters of that system. Quantization occurs when deviation is amplified to the point where no deviation-correcting mechanism can prevent the rupturing of the basic systemic framework—in other words, when the latter can no longer contain and canalize the energies and thrust which have been generated. The overall result alters the relationship of that system with its environment, creating new spatiotemporal, structural, and functional boundaries. In short, the system is transformed to a new level of internal organization and environmental integration.

Systems models have tended to be conceptualized "horizontally," that is, they are designed to examine either a single system—be it geomorphological, economic, or political—or, again, systems transacting with one another in a similar stage of structural development and behavior. But we want also to be able to include "vertical" shifts of societal organization so as to recognize increasing complexity and heterogeneity (see also Chapter 10 by Maruyama). Specifically, we seek a model which employs systems concepts to explain the circumstances in which quantization can occur from one societal stage to another. Hence, we are concerned to account for both (a) systemic levels of sociocultural organization, and (b) cybernetic processes that demonstrate (i) systemic self-stabilization within a given organizational level, and (ii) systemic transformation so as to result in a sociocultural quantum across an environmental frontier.

The perspective adopted in Figure 8.1 is that of the man–environment nexus which remains a planetary invariant. Viewed horizontally, each level depicts schematically transaction occurring among various environmental and societal factors. It is a geosocietal model inasmuch as we have broadly correlated levels of technological and governmental organization with stages and dimensions of environmental control. Viewed vertically, these stages of progressive overall environmental occupance assume a geometric sequence: point–line–plane–volume as mankind's control capabilities increase (within a global context). As the "Properties" column attests, this sequence also exemplifies the principle of integrative levels in action: each such level builds upon the properties and societal experiences of the level(s) below and

ISBN 0-201-03438-7/0-201-03439-5pbk.

SOCIETAL LEVEL	SYSTEM OF ENVIRONMENTAL CONTROL		PROPERTIES	EMERGENT QUALITIES				
	EXPLETED SPACE	IMPLETED SPACE		TECHNOLOGY	SCIENCE	TRANSPORTATION	COMMUNICATIONS	GOVERNMENT
S_5	Three-dimensional (extra-terrestrial)	Megalopolis ("Ecumenopolis")	BELOW +	Electrical-nuclear energy Automation Cybernetics	Einsteinian relativity Quantum mechanics Systems theory	Supra-surface: inner space systems Outer space explorations Surface systems Sub-surface vehicles	Electronic transmission (simultaneity throughout expleted space)	"Ecumenocracy" (Supra-national polities) Multi-level transaction Sovereignty invested in global mankind
S_4	Two-dimensional (oceans, continents)	City	BELOW +	Transformation of energy (steam) Machine technology Mass production	"Greek miracle" Scientific method Newtonian world-view	Maritime technology and navigation Thalassic and oceanic networks Highway networks Railroad technology	Mechanical transmission (printing) Alphabet	National state system Emergence of democracy Sovereignty of state (as primary actor)
S_3	One-dimensional (riverine societies)	Town	BELOW +	Non-biological prime movers (wind, water) Metal tools Continuous rotary motion (wheel) Irrigation technics	Mathematics Astronomy	Sailboats Riverine transport Wheeled vehicles Intra- and inter-urban roads	Writing	Ancient bureaucratic empires Theocratic polities Sovereignty of god-kings
S_2	Particulated Universal (sedentary)	Village	BELOW +	Animal energy Domestication of plants and animals Polished stone tools Spinning Pottery	Neolithic proto-science	Animal transport Paths, village routes Neolithic seafaring	Ideograms	Biological-territorial nexus Tribal level of organization and decision-making
S_1	Undifferentiated Universal (Nomadic)	Cave/tent (intraterrestrial)		Human energy Control of fire Stone and bone tools Partial rotary action		Human transport Sleds Dug-outs, canoes	Pictograms	Biological nexus (family, hunting band, clan)

FIGURE 8.1. Levels of organization.

ISBN 0-201-03438-7/0-201-03439-5pbk.

in turn contributes its own "emergent qualities," which take the form of new technologies and societal structures, accompanied by new apperceptions of the man–environment relationship. We can discern progressive developments in complexity and heterogeneity (although in any one historical situation a different, or even contrary, experience may occur).

In summary, Figure 8.1 provides a grid, showing both societal–environmental stabilization when viewed horizontally (process in planetary *space*-time) and societal–environmental quantization when examined vertically (process in planetary space-*time*). Mankind's overall experience has been to expand anthropogeographic perimeters concomitantly with its accelerative contraction of temporal sequences associated with new stages of environmental control, that is, from S_1—the food-gathering level of environmental control, through S_2—the food-producing level—and sequentially to S_5—our current developing level of geosocietal experience and organization.

The relation of species other than man to their environment is determined primarily by Darwinian, genetically coded mechanisms, so that the evolutionary process at the subhominid level can be described as *adaptive* correlation with the physical environment—because although the overall process is mutagenic and open-ended, and hence exhibits positive forms of feedback, negative feedback mechanisms dominate in the maintenance of individual species and their members. (In terms of Jantsch's "learning hierarchy," discussed in Chapter 3, we are dealing here with the *functional learning* mode that is characteristic of the whole bioorganismic world and is central to *phylogenesis.*)

Conversely, organisms with sensory-cognitive circuits are at the stage of *manipulative* correlation with the physical environment to the extent that they possess deviation-amplifying capabilities. It is by "man the toolmaker" that the equilibrating process shifts progressively from a reactively adaptive to an actively manipulative role. (This is akin to Jantsch's *conscious learning* mode which has "reformed ecological organization and made it a matter of conscious design" and which is central to *sociogenesis.*) Hence our model recognizes the crucial function of technology in the development and transformation of sociocultural systems from Paleolithic times to the present. In other words, science and technology—which we designate *material technics* (t_m)—largely serve as positive feedback processes. Concomitantly, we recognize the role of societal institutions and mores to maintain continuity and persistence in any sociocultural system. In this respect, therefore, *societal technics* (t_s) function as negative feedback processes to ensure overall stability and societal invariance under technological and environmental transformations.

ISBN 0-201-03438-7/0-201-03439-5pbk.

We can now diagram a systems metamodel which (a) accounts for (i) biospheric and (ii) sociocultural inputs from the total environment; (b) recognizes the given sociocultural system as (i) converter, (ii) within-inputs-generator (i.e., inputs generated within the system itself), and (iii) comprising numerous subsystems (including, e.g., the political); and (c) relates its outputs—as material and societal technics—to positive and negative forms of feedback.

Figure 8.2 indicates both how material and societal technics interact and, depending upon the state of the system vis-a-vis its environment, how they can combine so as to result in systemic self-stabilization or, alternatively, in systemic transformation. We shall designate the first systemic process "Cybernetics I"; the second "Cybernetics II." Figure 8.2 also aims to show that Cybernetics I, comprising net negative

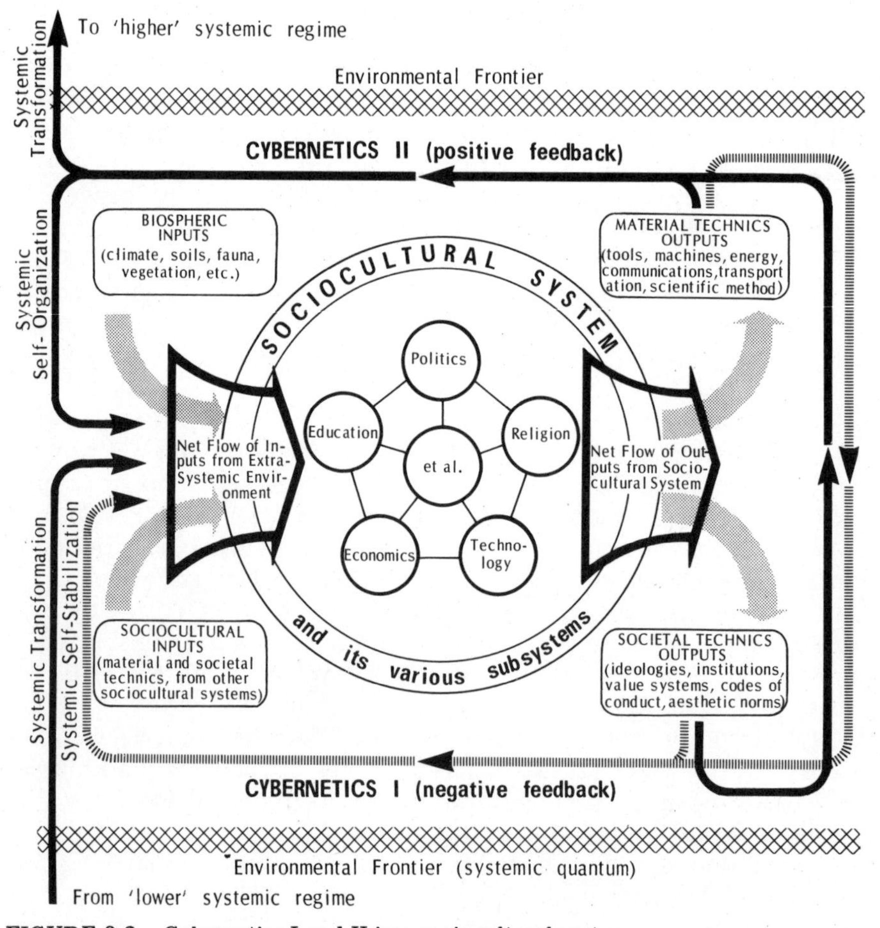

FIGURE 8.2. Cybernetics I and II in a sociocultural system.

ISBN 0-201-03438-7/0-201-03439-5pbk.

feedback processes, acts to stabilize a given sociocultural system within its environment. In contradistinction, the dominant positive feedback processes comprising Cybernetics II can (i) increase the system's negentropy and information gain, and thereby also increase its environmental control capability so as to actualize the existing potential within the system–environment nexus and/or (ii) enable the system's outputs (in the form of material and societal technics) to cross the permeable frontiers separating one environment from another and so quantize to a new level of societal organization.

Examples of Cybernetics I are found in subhominid societies where Darwinian mechanisms are fully operative; again in mature or senescent sociocultural systems in which the available material technics have achieved their maximal environmental control capability and reached steady state. As an example of Cybernetics II (systemic self-organization), we might choose lithic man's advancement into the high latitudes. It was made possible by control of fire (energy production) and invention of progressively efficient and specialized tools (such as microliths). By attaining maximal use of this technology, the Eskimos could survive beyond the tree line and maintain a viable symbiosis with an austere (i.e., low-energy) physical environment. However, since the latter sets constraints on expansion and control of the biosphere, negative feedback mechanisms became dominant, resulting in overall societal stabilization (Cybernetics I)—so that society was to remain at S_1 (in Figure 8.1), at least until the intrusion of an alien, more advanced technology.

The second type of Cybernetics II (systemic transformation) is exemplified in preliterate Southwest Asia where material technics acquired at S_1 were augmented by technological withinputs, that is, domestication of the ancestors of wheat and barley and of certain wild animal species. The ensuing "Neolithic Revolution" (Gordon Childe's term) transformed the man–environment relationship and quantized it to S_2 by extending Neolithic society's environmental control capability and creating new settlement patterns, new societal organization, and a more complex division of labor, while an increased and more reliable food supply enabled a larger population to be supported. However, at this stage of incipient agriculture, limited water resources (as at Jericho and Jarmo) set rigid limits upon growth, so that eventually steady state (Cybernetics I) resulted.

A further systemic quantum occurred when these material technics were applied to a different kind of environment. The transplanting of S_2 agricultural innovations to the rich bottom lands of the Nile, Tigris-Euphrates, Indus, and Huang-ho yielded a hundredfold increase in food harvests, making possible a "social surplus" that unprecedentedly increased population numbers and densities, and freed many persons to

ISBN 0-201-03438-7/0-201-03439-5pbk.

work in occupations and localities at some remove from the fields. Hence the rise of towns, accompanied by more complex governmental and administrative structures, hieratic elites, and so on. In short, the "Urban Revolution"—S_3—describes a systemic quantization-cum-environmental-transformation in which human control eventually covers an entire river valley and raises technological and societal activities to a new negentropic level and complexity, resulting in the first "civilizations." With this hydraulic technology, control of the river eventually reaches a plateau of stabilization because of the constraints inherent in the surrounding environment. Thence Cybernetics II yields once more to Cybernetics I, the latter reinforced by societal technics which maintain systemic stability for millennia in these archaic riverine civilizations, until they are in turn disturbed—destabilized—by sociocultural systems possessing still more sophisticated technologies and more potent environmental control capabilities.

Thus, if space permitted, we could utilize our metamodel to analyze societal complexification and quantization, with their attending environmental transformations, at S_4 and S_5 in turn. At this juncture, however, we want to turn to a major theme in this volume and employ our systems approach in an effort to come to grips with the question posed in the introduction to this chapter.

3. ORDER THROUGH FLUCTUATION IN HUMAN SOCIETIES

As discussed by Prigogine in Chapter 5 and by others in this volume, living organisms and sociocultural systems alike may be regarded as partially open systems which interact with their respective environments and are themselves in states of nonequilibrium. The new field of nonequilibrium thermodynamics has postulated the principle of "order through fluctuation" whereby such systems, being partially open to the inflow of energy (information) and/or matter, develop instabilities which, however, do *not* lead to random behavior (even if the initiating fluctuations as such are random). Instead, they tend to drive the system to a new dynamic regime which may correspond to a new state of complexity. Nonequilibrium systems are characterized by a high degree of energy exchange with the environment (and can therefore be termed dissipative structures). Thus, nonequilibrium dynamics is moving toward a theory of self-organization of processes and structures, applicable not only to the physical but to the biological and social domains as well.

Consistent with this principle of order through fluctuation, we have thus far been applying systems concepts to the historical record in

ISBN 0-201-03438-7/0-201-03439-5pbk.

order to demonstrate an overall, global societal evolution marked by periodic self-organization and (on balance) progressive complexification. At this juncture, let us briefly analyze the principle of order through fluctuation in terms of the following properties of a system: (1) stability; (2) resilience (persistence); (3) complexity; and (4) quantization (discontinuity); and then relate these properties to human societies.

In Chapter 4, Holling defines *stability* as a system's ability to return to an equilibrium state after a temporary disturbance, and *resilience* as its ability to absorb change and to persist (i.e., to maintain structural integrity in its development or management). He points out that a system can be very resilient and still fluctuate greatly, that is, possess low stability. He also states that the more homogeneous the environment in space and time, the more likely is the system to have low fluctuation and low resilience. (Conversely, we can infer that the more heterogeneous the environment, the more likely is the system to exhibit highly fluctuating modes and high resilience.)

In the human world, an example marked by unparalleled persistence, minimal impact upon the environment, and the constraints of societal tradition is Paleolithic society. Lithic communities tended to be structurally simple and environmentally homogeneous, and to experience low fluctuations; that is, they appeared to the outside as highly stable. But Holling's suggestion that systems with low fluctuations are also more likely to have low resilience may help account for the well-documented phenomenon that once primitive (simpler) societies are subjected to major impacts from more complex societies—to significant external fluctuations—they can rapidly become dysfunctional to the point of systemic collapse.

What, then, is the relation between "stability" and "complexity" (which we may define as the number of linkages within a system or, again, between a system and its environment)? A central theme in population ecology is that increased trophic (nutritional) web complexity leads to increased community stability; that is, the greater the number of links and alternative pathways in the web, the greater the chance of absorbing environmental shocks, and thus of damping down incipient oscillations.

Relevant here is also the time factor as it pertains to natural ecosystems and, again, to human societal systems—and, a propos of the latter, to the very different time scales of Stone-Age and modern societies. In comparison with natural ecosystems, all human systems are both less stable and more complex (inasmuch as they build upon the former in turn). Yet compared with their modern successors, Paleolithic societies (S_1 in Figure 8.1) possess marked structural simplicity and temporal stability, whereas our own society is distinguished by unparalleled

ISBN 0-201-03438-7/0-201-03439-5pbk.

complexity—it has traded stability for richness in adaptability (Laszlo, 1972).

Contemporary society is characterized by a rapid dissemination of information compared with primitive societies, and the resulting flexible coupling between subsystems leads to conditions for metastability which become more pronounced with higher complexity (see Chapter 5 by Prigogine). Because of the marked coupling of its subsystems, a modern societal system can usually sustain relatively large fluctuations (deviations from the norm) and still maintain structural integrity. As these fluctuations attain critical size, however, the system verges on quantum transition, which occurs if the fluctuations rupture the parameters of the system itself or, again, those which couple the system with its given geosocietal environment. This occurs when a society alters its material and social technics, formulates new goals, and alters its environment. At one stage of systemic organization, the societal system utilizes negative feedback for self-stabilization (Cybernetics I); in a second time frame, a technological innovation, the introduction of a new type of government, or the formulation of new sociopolitical goals may act as a catalyst to cross the "threshold level" which nullifies the overall effects of the system's existing negative feedback processes. As a result, the system is reorganized internally, and shifts to a new level of integration with its environment, which in turn is thereby altered (Cybernetics II).

If, as both empirical evidence and conceptual innovation seem to attest, we appear to be in the midst of a rapid and accelerating societal quantum shift from S_4 to S_5 (in Figure 8.1), we require a synthesis of models conceptualizing dynamic and metastable forms of societal life in order to account for our contemporary paradigmatic and behavioral shifts.

4. TOWARD A NEW SOCIETAL REGIME

To this point, we have been constructing a system model that is focused upon a *single* sociocultural system and its various subsystems (including, e.g., the political subsystem which takes the form of a "polity" of some kind in the geopolitical environment). But few if any societies or polities have ever existed in spatial isolation. In nonsystemic approaches to political science and political geography alike, the structure, power inventory, and environmental capabilities of polity A are compared with polities B, C, . . . , N. Thus, when one state as "primary actor" on the world stage is set against another primary actor, the resulting model

ISBN 0-201-03438-7/0-201-03439-5pbk.

is a "balance of power" mechanism. However, this traditional approach is *summative*, that is, it adds up the properties of polities viewed separately as though the resulting total will be able thereby to account for the actors' individual and net behavior alike. But a systems approach (in which the complex is seen as *constitutive*) tells us that perceptions, decisions, and resulting geopolitical outputs are indelibly altered by the relationships resulting from the interactions among polities A, B, . . . , N. Moreover, as we know all too well in today's world of planetary ecological disequilibria, these two or more polities do not exist in some isolated environment. All other political entities are also involved, together with the human and material resources comprising the terrestrial biosphere. Hence, a systems approach to international relations conceptualizes it within a global "field" of continuous transaction.

At this juncture, we might construct a simple systems geopolitical model to show the types of transaction occurring between two or more polities of comparable societal organization, such as contemporary nation-states (which share juridical isomorphisms). This model seeks to demonstrate not only that one polity's perceptions, decisions, and geopolitical implementations are transacting with those of the other, but that their joint feedback loops are also connected with biospheric and sociocultural processes occurring in the extrasystemic environment—that is, the environment enveloping the *de facto* "system" which the two polities' reciprocating perceptions, goals, and activities have thereby constructed (Figure 8.3).

As we have noted, our sociocultural metamodel (Figure 8.2) sought to show that process also occurs vertically, that is, from one level of systemic organization to another. System principles of integrative levels, isomorphy, stabilization, and quantization in conjunction go far to explain not only systemic reorganization and environmental transformation, but also the emergence of new properties or attributes, greater societal complexity, and increased environmental control at each succeeding level. The overall historical evolution of planetary society in its organizational-environmental framework can be broadly schematized in turn, as in Figure 8.4.

The thrust of our technology and its time–space telescoping capabilities, and the complexification of our societal structures, are hurtling us laterally across boundaried national space and vertically into new levels of systemic organization and integration. Moreover, the attainment of each systemic level brings with it a greater, horizon-receding perspective, and a new set of emergent properties and options. Hence we are led to the demise not of the territorial state per se but of the traditional geopolitical construct in which the physical environment and societal values and behavior, including those emotive attributes of the nation-

ISBN 0-201-03438-7/0-201-03439-5pbk.

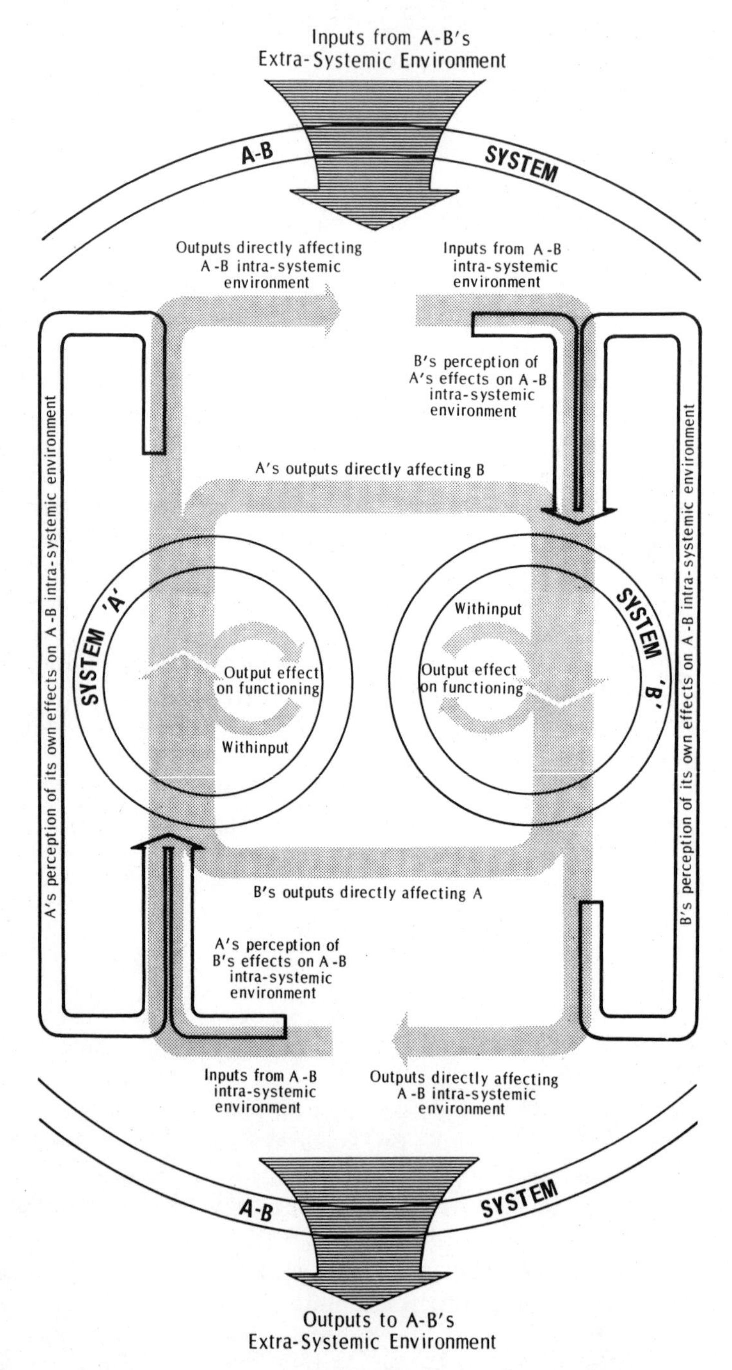

FIGURE 8.3. Model of two geopolitical systems in interaction.

ISBN 0-201-03438-7/0-201-03439-5pbk.

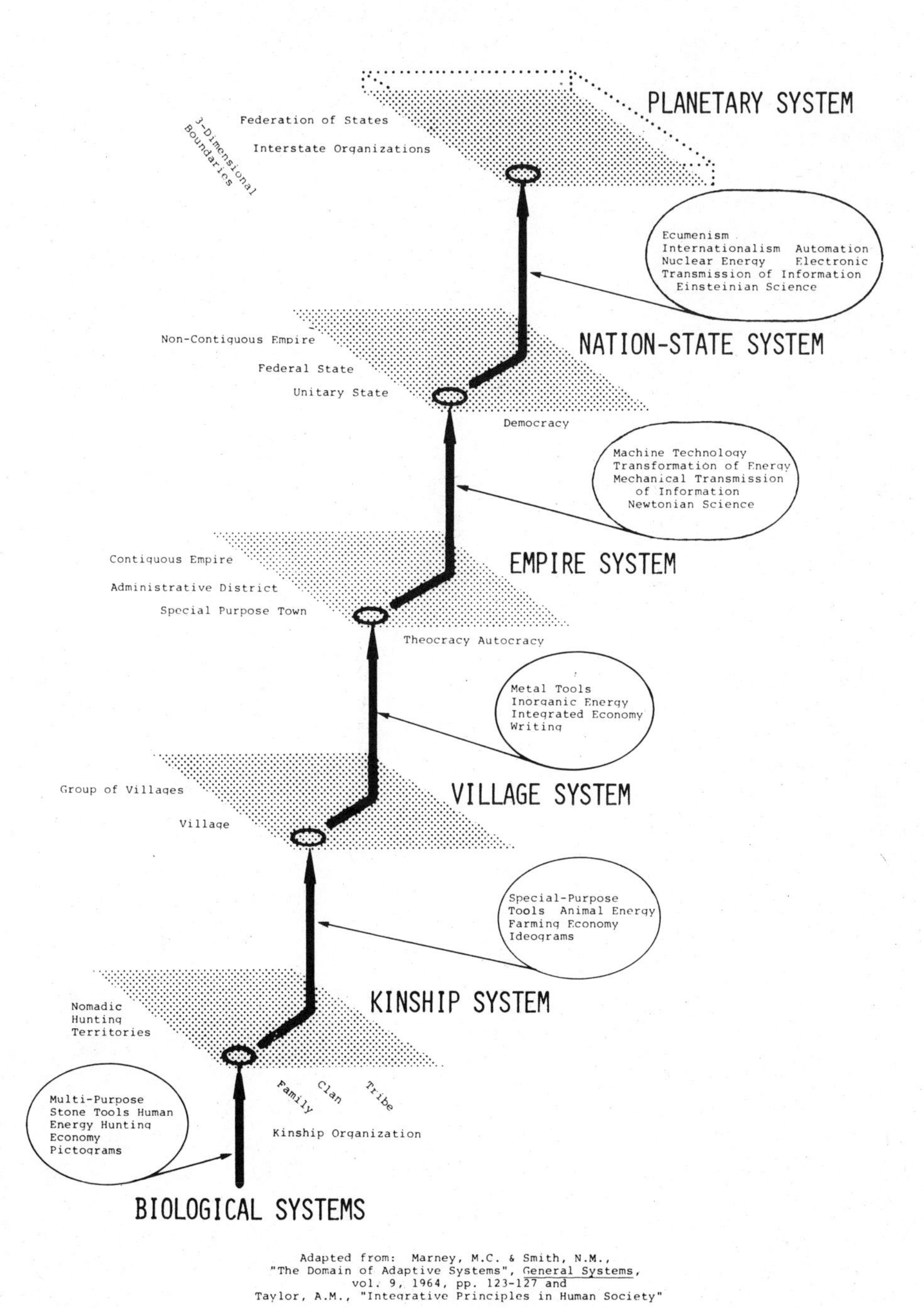

Adapted from: Marney, M.C. & Smith, N.M., "The Domain of Adaptive Systems", General Systems, vol. 9, 1964, pp. 123-127 and Taylor, A.M., "Integrative Principles in Human Society"

FIGURE 8.4. Emergent geopolitical systems levels.

ISBN 0-201-03438-7/0-201-03439-5pbk.

state such as "sovereignty" and "patriotism," are neatly compartmentalized within territorial boundaries. A systems approach rejects these simple exclusive alternatives in favor of multirelational orientations—each of us lives concurrently at more than one level of sociopolitical organization, municipal, state, national—and the thrust of twentieth-century material technics and the emergence of new societal technics require the evolution of human societies toward a planetary plateau of sociocultural organization and integration.

What we have to recognize is our shared involvement in a fundamental conceptual shift—a multirelational transformation from a nation-state paradigm progressively to a global construct comparable to the shift of perspective from the Mercator projection—rectilinear and emphasizing two-dimensional "flat" space—to, say, an orthographic projection, at once curvilinear and recognizing new spatial relationships. Perhaps the traditional concept of space itself requires to be assessed anew. The nation-state paradigm tended to view space as a void, an empty receptacle to contain pieces of "property," so that space was largely a matter of "place" and "location," and what lay beyond the property lines was either no-man's land or, alternatively, open to entrepreneurial "grabs" and exploitation. But a very different way of perceiving space is to regard it as a *plenum*—an ordering constituent of a macrocosmic system in which field forces are omnipresent and omnioperative, acting upon all material phenomena and maintaining a dynamic, energizing, as well as balancing, *field.* It is in the context of a plenum or field that we need to approach the ordering of our planetary and extraterrestrial spaces alike.

5. CONCLUSIONS

In conclusion, what we have attempted to construct is an evolutionary model of sociocultural (and geopolitical) process and structure, one which correlates specific levels of societal and political organization with characteristic stages of environmental control as expressed in terms of science, technology, and social institutions. Each new level in the ascent of societal evolution from tribal to global systems is first characterized by the generation of positive feedback processes so as to organize its key elements, followed by the dominance of negative feedbacks to assure the maintenance of societal organization at that specific level. This model is particularly concerned with understanding qualitative transformations—or mutations—within what Jantsch (1975) terms the "social mode of organization, if it is understood that several levels

ISBN 0-201-03438-7/0-201-03439-5pbk.

are in play simultaneously." He adds (1975, p. 279): "As a higher geopolitical level is dominated by positive feedback establishing a new dynamic regime, lower levels become coordinated by negative feedback. As national systems become more and more subject to homeostatic control in the presence of stringent constraints, regional systems and the emerging global system reach out with positive feedback, groping for their viable structure."

One final thought as we correlate our sociocultural model with the concepts of "stability" and "resilience": In his study of ecological systems, Holling (Chapter 4) points out that in the management of resources, the stability view emphasizes equilibrium and the maintenance of a predictable world. In contrast, the resilience view

> would emphasize the need to keep options open, the need to view events in a regional rather than a local context, and the need to emphasize heterogeneity. Flowing from this would be not the presumption of sufficient knowledge, but the recognition of our ignorance; not the assumption that future events are expected, but that they will be unexpected. The resilience framework can accommodate this shift of perspective, for it does not require a precise capacity to predict the future, but only a qualitative capacity to devise systems that can absorb and accommodate future events in whatever unexpected form they may take.

Such recognition is required for our contemporary stage in sociocultural evolution, which can be described as one of transformation. As in earlier transformational epochs, such a stage will be dominated by positive feedback processes, and must be prepared to accept high societal fluctuations and low stability, with an important innovation: Whereas in the past science and technology were the chief agents of positive feedback in transforming the man–environment relationship, while societal technics provided the negative feedback mechanisms to stabilize that relationship, perhaps what will be required in the years ahead is a progressive reversal of these traditional roles. Given our finite resources and the need to accommodate a still burgeoning planetary population, our science and technology must devise new methods for conserving those resources and stabilizing our global ecology, while we must also in turn give full rein to the creation of new social institutions and values to encourage heterogeneity and new states of awareness among our cultures (see also Chapters 10 by Maruyama and 11 by Markley). Indeed, what may be essential is the recognition of the need to engage in the *continuous* invention of new and appropriate societal technics to unlock the hitherto largely untapped creative potential of cultures and individuals alike in an evolutionary process that is always

ISBN 0-201-03438-7/0-201-03439-5pbk.

mutagenic and open-ended. Perhaps only in this way can we hope to "absorb and accommodate future events in whatever unexpected form they may take."

REFERENCES

Berghe, Pierre L. van den (1963). "Dialectic and Functionalism: Toward a Theoretical Synthesis," *Amer. Sociological Rev.*, 28, 695–705.

Demerath, N. J., III, and Peterson, R. A. (eds.). (1967). *System, Change, and Conflict.* New York: Free Press.

Frey, Dagobert (1956). "On the Problem of Symmetry in Art," *Studium Generale*, p. 276; quoted by Hermann Weyl in *The World of Mathematics* (James N. Newman, ed.), Vol. 1, p. 678. New York; Simon and Schuster.

Horowitz, Irving Louis (1962). "Consensus, Conflict, and Cooperation: A Sociological Inventory," *Social Forces*, 41, 177–188.

Jantsch, Erich (1975). *Design for Evolution: Self-Organization and Planning in the Life of Human Systems.* New York: Braziller.

Laszlo, Ervin (1972). *Introduction to Systems Philosophy: Toward a New Paradigm of Contemporary Thought.* New York: Gordon and Breach; Harper Torchbooks.

ISBN 0-201-03438-7/0-201-03439-5pbk.

9

Evolution of Scientific Method

Milton Marney and Paul F. Schmidt

1. INTRODUCTION

It will be generally admitted that the development of the historic agencies of man's correlation with his environment—the great institutional triad of religion, science, and art—depended utterly upon the power of significant symbols to render thought coherent, communicable, corrigible, and therefore capable of interlacing human individuals into a nexus of socialized creativity and control. A generalization which makes symbolization and cognition the foundational factors in cultural development amounts, further, to the proposal of a synoptic hypothesis: that the emergence of successively dominant patterns of sociocultural organization is interdependent with *noetic evolution*, with the emergence of successively dominant modes in the formulation of concepts, theories, and intellectual methods.

If we are ever to get within reach of the convoluted significance of science as an institution, we must first comprehend the significance of the omnipresent theme of a cognitive modality providing linkage and support for the whole institutional triad of which science is a part. We must succeed in constructing a rationale by which sophisticated intellectual achievements of science become recognizable as extensions of the elemental cognitive-semiotic complex underlying all cultural development. Figure 9.1. attempts to sketch the emergent structure of the cognitive-semiotic-social complex at three stages of societal evolution.

Science will then represent simply the proliferation of a particular version of the cognitive modality. It will be identified with an evolutionary lineage, a sequence of emergent but nonetheless connectable *ways of thinking*; and this lineage will be punctuated by appearances of prototypes of rational thought. At this remove, historical instances of scientific advance will be distinctively marked by the advent of adaptive modifications of the cognitive modality itself, that is, by modes of thought that serve the function of cognition with increasing adequacy and effectiveness.

ISBN 0-201-03438-7/0-201-03439-5pbk.

Erich Jantsch and Conrad H. Waddington (eds.), *Evolution and Consciousness: Human Systems in Transition.*

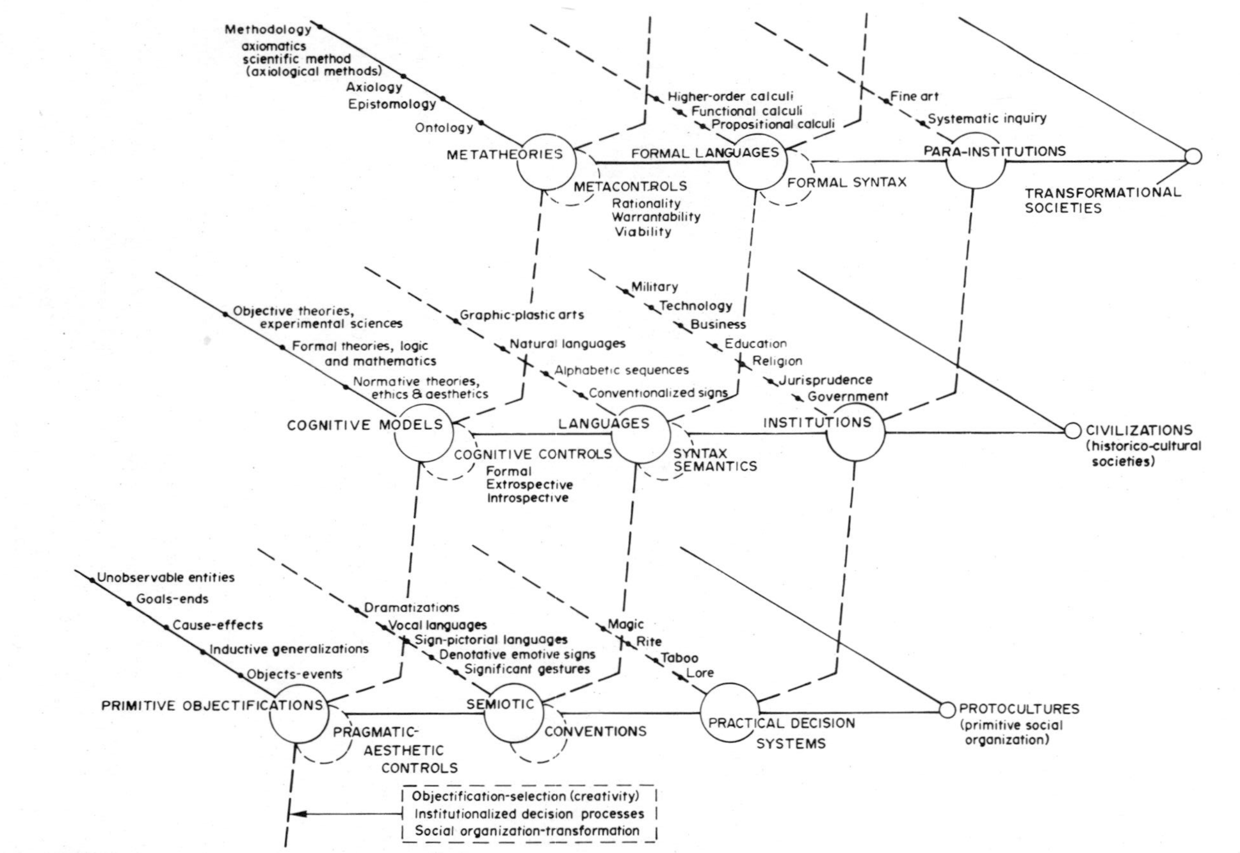

FIGURE 9.1. Emergent cognitive–semiotic–social structures.

ISBN 0-201-03438-7/0-201-03439-5pbk.

2. SUCCESSIVE PROTOTYPES OF SCIENTIFIC INQUIRY

Three principal rational prototypes, each one featuring a basic modification of its predecessor, apparently have directed the course of Western science in this sense. These we may term (1) the axiomatic, (2) the empirical, and (3) the constructural prototype. A fourth one—the normative prototype—is emerging in our day.

Axiomatic Prototype

Upon a confused welter of uncontrolled prescientific speculation, the *Organon* of Aristotle (384–322 B.C.) imposed an elegant organization of scientific procedure which was implemented classically in Euclid's *Elements of Geometry* (around 270 B.C.). The foundations of the axiomatic mode were laid on the following philosophical assumptions:

1. that serious and persistent reflection must ultimately result in the intuitive apprehension of certain "most general" propositions (*archai*, or axioms) undeniable in character and therefore acceptable to all rational investigators;
2. that unique definitions and self-evident premises were attainable by agreement of all persons trained in a given subject;
3. that valid procedures of deductive reasoning operating on definitions, premises, and axioms would produce necessary conclusions as theorems which were true, independent of experience, yet universally applicable to the physical world; and
4. that the systematic development of the entire complex of theorems possible on this basis would comprise a body of universal knowledge.

By the time of Copernicus (1473–1543), a restricted but influential community existed for which the conception of science involved a thorough-going incorporation of a lost maxim of Aristotle: that scientific knowledge must comport with observations. Yet it is clear that the addition of Socratic hypotheses and Aristotelian empiricism represented an insufficient modification of the axiomatic method. In the great debate that arose concerning the acceptability of Copernican versus Ptolemaic astronomical theories, a disconcerting realization appeared. Science had encountered a prime example of the fact that two different hypotheses, logically quite incompatible, may be equally confirmed by experience. True enough, there existed certain extralogical considerations which afforded criteria for a choice between alternatives. By virtue of its superior elegance the Copernican theory seemed preferable; but the demand at the time was for an intellectual basis for a decision as to which was the truly applicable theory.

ISBN 0-201-03438-7/0-201-03439-5pbk.

The identical obstruction on which a purely rationalistic axiomatic mode had foundered now confronted the new empirical science, namely, the necessity of achieving some definitive test leading to the resolution of conflicting claims. Confusion had merely been compounded by the additional weight of experimental verification, since it might be used to justify either of two incompatible hypotheses. It was nevertheless a sharpened version of empiricism which was to lead to a successful modification of scientific method.

Empirical Prototype

Systematizing ideas and procedures originating with Kepler, Galileo, and Bacon during a century of precursory work, Isaac Newton's *Principia Mathematica* (1687) struck a new balance between the roles of inductive and deductive procedures of inference. On the one hand, empirical observation—under rigorous control of precise measurement and careful generalization—was instituted as a primary directive to the formulation of premises. The working hypotheses of science need no longer be limited to principles derived by purely reflective reasoning (intuition). On the other hand, mathematical disciplines and procedures carried over from the prior axiomatic model were to be employed in derivation of theoretical consequences, some of them perhaps unforeseen; and in this promising event, experiments were to be designed specifically to test a given theory for correspondence of its consequences (predictions) with facts obtained by experimentation. The principal assumptions involved in this version of scientific method were that:

1. uniformities of nature as "natural laws" were extractable from observations of phenomena, presumably via creative insight;
2. basic postulates could specify factual relations among primitive (undefined) quantitative concepts as abstract descriptions of the real character of the natural world;
3. contemporary axiomatic systems of mathematics—assumed to be universally valid and applicable to the physical world—were adequate for derivation of predicted observations; and
4. the confirmation of a theory should result from its correspondence with relevant experimental evidence.

Euclidean geometry had furnished the conceptual framework which Newton utilized. The development of a staggering collection of distinct, internally consistent, non-Euclidean geometries (following a format provided by Georg Riemann around 1850) proved more than a little

ISBN 0-201-03438-7/0-201-03439-5pbk.

disconcerting to the view that Euclid's axioms and postulates were self-evident truths necessarily applicable to all natural forms and processes. The characteristic conclusion of the nineteenth century was, however, that this proliferation of geometries involved purely abstract creations, of interest only to formal science, while Newtonian physics indicated that physical space was in fact Euclidean.

Under two great themes, mechanics and electrodynamics, classical physics claimed comprehension of a tremendous scope of physical phenomena—a picture complete but for details, so it might have been regarded. But there were troublesome aspects in the apparent incoherence of the two divisions. Newton's laws of motion and gravitation seemed irreconcilable with equations for the propagation of electromagnetic energy brilliantly formulated by James Clerk Maxwell in his *Electricity and Magnetism* (1873). The breakthrough contribution of Albert Einstein, in *General Theory of Relativity* (1916), proffered a basis for merging these apparently disparate fields. Appropriating (from Lobachewski) one of the non-Euclidean geometries which had seemed so thoroughly counterintuitive, Einstein succeeded in formulating the fundamental equations of an analytical mechanics in which the previously intractable distinction between gravitational and electromagnetic forces no longer figured. In this novel conceptual format, the distinct categories of space, time, matter, and energy were withdrawn and replaced by multiple components of a single formal entity: a unitary *field* characterized by a metrical geometry of the Riemannian type. This new four-dimensional theoretical model adequately accounted for the traditionally significant phenomena in both the classical fields of mechanics and electrodynamics.

A third version of scientific thought—the constructural prototype—had begun to emerge as a conception of inquiry in which any arbitrary formal scheme whatever might be utilized for the organization and explanation of selected domains of experience—subject to strict contraints of logical consistency and legitimate interpretation.

Constructural Prototype

In the form of this conception, subsequent investigators in the twentieth century received a legacy of such proportions that tremendous effort had to be expended merely in comprehending and exploiting its potential. A notable revolution of modern physics, beginning in the 1920s, featured exploitation; and the project of general assessment still absorbs the interests of contemporary philosophy of science. The following commitments outline the constructural mode of inquiry as it stood at roughly mid-century.

ISBN 0-201-03438-7/0-201-03439-5pbk.

Factual knowledge was to be attained by the combined employment of two independent divisions of inquiry:

1. *formal* science—the domain of axiomatic systems—in which conventional, logical schemata (not self-evident propositions) were to be devised under the control of deductive logic; and
2. *empirical* science—the domain of explanatory object-theories—in which a selected formal schema, interpreted in terms of observable measures, was to be manipulated in order to elucidate the consequences of a given theoretical model and to test the correspondence of predictions with experimental evidence.

Abstract mathematical schemata disclosed nothing definitive about the character of nature; they merely presented conventional internally consistent *a priori* forms devoid of content until assigned an experiential interpretation.

The elemental procedure of inquiry consisted in:

1. construction of theoretical models relevant to some specific domain of experience;
2. testing by experimentation designed to exhibit possible inadequacies of the theory; and
3. modification or reformulation of a theoretical model which had been thus disconfirmed. It was generally supposed that exceptional events could disprove a theory, although in fields where only statistical inference was possible the disposition of a theory clearly rested with some arbitrary rejection principle.

The quest for certainty was relinquished; scientific knowledge was *a posteriori* (dependent on the outcome of experience) and probabilistic in all fields of natural science. Such knowledge was subject to pragmatic and aesthetic controls designed to ensure effectiveness for purposes of prediction, explanation, and control. It must ultimately comprise a coherent philosophy of nature, and it must prove satisfactory under demands for simplicity, elegance, and comprehensiveness.

It is therefore understandable that an assumption prevailing from early modern science should have been widely accepted and carried over in this latest modification of scientific method. This assumption was that straightforward exploitation of the mode which led to successful prediction and explanation of mechanistic systems could be expected to provide an adequate basis for comparable successes in the social and life sciences. Persistent attempts to extend the success of exact objective inquiry into the arena of the social and life sciences, however, have encountered intractable problems when confronted with the modifiability of characteristic response (the *adaptivity*) of organismic systems. It is not to be denied that strict extension of the method of the exact sciences has led to impressive accomplishments in both micro-

ISBN 0-201-03438-7/0-201-03439-5pbk.

behavioral studies of molecular biology and physiological psychology and in macrobehavioral modeling for economics and ecology. The point is simply that these attainments have been limited either to the consideration of elementary subsystems or to abstract reductions of the whole organisms (organizations) that are of ultimate interest in biology, sociology, psychology, anthropology, ecology, and economics.

Contemporary Modification: Normative Prototype

A hierarchical range of interactions affects the response of any complex living system. As a consequence, the requirement of the so-called system sciences is for total-system models adequately reflecting the intricate connectedness of multilevel, multigoal organizations in which positive and negative feedback processes give rise to alternative decision modes which may be termed as follows:

1. *reaction:* response determined by a program of habitual behavior;
2. *reprogramming:* modification of existing operational programs or allocations of resources;
3. *renormalization:* resetting of established norms to secure either (a) feasability by relaxation, or (b) improved control via more demanding criteria;
4. *reorganization:* redesign of structural, functional, or valuative features of a system; and
5. *reobjectification:* reconstruction of previous conceptions of the external world, or (rarely) the principles presumed to hold over interaction (see also Chapter 11 by Markley).

A "normative-theoretic" mode of inquiry emerged as a modification of the constructural prototype. According to this new way of thinking, the objects of scientific interest in general are conceived as:

1. value-sensitive organizations, presupposing change in structure or behavior at every level of analysis to be associated with selective responses of subsystems;
2. coordinated by a protocol or regime (a synthesizing principle intrinsic to the design of the system as a whole) and specifying a system-wide coupling of decision criteria;
3. tending to drive adaptive behavior toward the extremalization of a holistic value function (such as, depending on the complexity of the system in question: minimization of stress; maximization of effectiveness in purposive goal-seeking; optimal allocation of resources; optimization of strategy, policy, and organization design in order to ensure the continuity of exchange functions with the environment; maximal realization of the potential of a given system configuration; or stabilization of a supreme value measure associated with optimal tradeoff of conflicting values such as maximal freedom, optimal control, and maximal scope); and

ISBN 0-201-03438-7/0-201-03439-5pbk.

4. encompassing all of the natural norms specific to a given individual selective system enmeshed in a process of evolutionary selection—whether physical, biological, sociocultural, or conceptual.

In principle, the range of consistent application for *normative-hierarchic* restructuring therefore admits of a unification of the previously disparate major domains of mechanistic versus organismic systems. An alternative prototype, *ecosystems modeling* (see also Chapters 4 by Holling and 5 by Prigogine), constitutes a first-order approach to a global system which has been purposefully *under*specified by representation merely as a system of entropic, economic, and social *transactions* among a collection of mutually adaptive systems. An impressive number of distinct disciplines now tend to address problems which appear to be associated, in fact, with completely *general* features found in every instance of adaptive systems weakly coupled by transactions, as distinct from the strong normative coupling typical of unitary organizations as command-control, or organismic, systems. Thus, ecosystems modeling has been applied to the natural ecology, managed natural ecologies (see also Chapter 4 by Holling), human settlements, economic systems, the technosphere, sociopolitical systems (see Chapter 8 by Taylor), and information systems and cognitive models (see also Chapter 10 by Maruyama). The observed structural, dynamic, and regulatory properties of ecosystems in these vastly different domains exhibit striking analogies (see Tables 9.1–9.3, which also emphasize the relations between some of the core notions used in this volume).

The normative approach constitutes a move toward *ending the exclusion of values.* In earlier work related to these issues (Marney and Smith, 1972) it has been shown that

1. the epistemological status of "value" as a formal construct is in no way different from that of comparable constructs (e.g., momentum);
2. mathematical structures which admit of interpretation in terms of value concepts are already reasonably developed in first-order perturbation theory, contemporary mathematics of optimization and calculus of variations, theory of stochastic indefinite processes, and theory of dissipative structures (see also Chapter 5 by Prigogine);
3. for every objective representation of a behavioral system in terms of deterministic causal relations or stochastic processes, there exists, in principle, a complementary normative representation in the form of a mathematical dynamic program for the extremalization of a suitably formed value function subject to constraints; and
4. cybernetic characteristics of the finite cognitive agent entail the relativity and reductivity of all conceptual representations and, hence, the necessity for *complementary* representations of substantive versus valuative determinants of systems behavior.

ISBN 0-201-03438-7/0-201-03439-5pbk.

ISBN 0-201-03438-7/0-201-03439-5pbk.

TABLE 9.1 Observed Ecosystem Properties: Structural[a]

	DISCIPLINES						
PROPERTIES	*Natural systems ecology*	*Thermodynamics and chemical evolution*	*Information theory and automata theory*	*Human geography and settlement pattern*	*Reliability analysis and control theory*	*Economics and technosphere*	*Physiology and psycho-physiology*
Hierarchical	Food Chain	Locally stable configuration (Wheeler)		Urban hierarchy		Industrial complexes	Organismic structure
Rank-size rule	Species per genera		Word frequency (Zipf)	Distribution, City sizes		Corporate assets	Allometric growth
			Articles per journal (Bradford)			Income distribution (Pareto)	
Sigmoid profile on environmental gradients	Species packing on uniform clines	Diffusion barriers		Influence of neighboring cities		Market areas, competition	
Self-organizing	Reproduction, order through fluctuation (Prigogine)	Nonequilibrium thermodynamics, dissipative structures (Prigogine)	Self-reproducing automata	Urban development, symbiotization of heterogeneity (Maruyama)	Ultrastable control	Economic and social development (Dunn)	Metabolism
Structural code	Genetics	Auto-catalyst	Letters, words, codes	Cultural economic functions		Technology Skills	Phonemes in spoken language

[a] Adapted from Marney and Anthony, 1974.

TABLE 9.2 Observed Ecosystem Properties: Dynamic[a]

	DISCIPLINES						
PROPERTIES	*Systems ecology*	*Thermodynamics and chemical evolution*	*Information theory and automata theory*	*Human geography and settlement pattern*	*Reliability analysis and control theory*	*Economics and technosphere*	*Physiology and psycho-physiology*
Logistic substitution	Emergent species	Combustion processes		Regional growth		Technological substitution	
Limit cycles	Predator-prey cycles	Autocatalytic reactions Chemical cycles in nonequilibrium process				Business cycles	Cell chemistry
System breaks	Epidemics	Phase change Explosion	Signal lost in noise	Collapse of a civilization	Catastrophic positive feedback Loss of power	Depressions, ecospasm	Death
Complexity, fluctuations, and stability	Resilience (Holling)	Stochastic models Metastability	Channel capacity	Space as plenum (Taylor)		Metastability /Resilience (Prigogine, Holling)	Sickness, self-healing, psychosomatic effects

[a] Adapted from Marney and Anthony, 1974.

ISBN 0-201-03438-7/0-201-03439-5pbk.

ISBN 0-201-03438-7/0-201-03439-5pbk.

TABLE 9.3 Observed Ecosystem Properties: Regulatory[a]

	DISCIPLINES						
PROPERTIES	*Systems ecology*	*Thermodynamics and chemical evolution*	*Information theory and automata theory*	*Human geography and settlement pattern*	*Reliability analysis and control theory*	*Economics and technosphere*	*Physiology and psycho-physiology*
Redundancy and diversity of function	Multiple species and natural selection		Error-correcting codes	Division of labor	Multiple paths	Balanced portfolio	Brain function
Reponse times	Reproduction and growth rates	Reaction rates	Encoding/ decoding Transmission delays Computer time sharing	Migration of business and population	Feedback stability	Spreading q of innovation	Learning theory
Interaction network	Niche competition	Reaction "wave-length" Critical size	Communication network	System of cities Transportation Communication		Contact intensive jobs	Nerve network
Logarithmic response of sensors			Information encoding			Exponential growth	Visual auditory response

[a] Adapted from Marney and Anthony, 1974.

In sharp departure from the view of heuristic system modelers, who hold that the specification of value parameters of a human system can readily be stipulated, or determined by quick iteration of trial commitments, the normative-theoretic approach to modeling dictates a counterproposition: that the warranting of values does not admit in principle of a "crucial experiment" type of testing. There is no royal road to tested value commitments. In the field of value theory, *vindication* is the analogue of validation and confirmation (see also Chapter 3 by Jantsch).

The emergence of novel modes of scientific thought does not end here. Our account comes to a halt only because we are now caught up in action at the edge of advance.

3. EMERGENT FEATURES OF SCIENTIFIC ADVANCE

On the basis of successive prototypes of scientific inquiry, as discussed in the preceding section, we may turn with good conscience to the identification of emergent features of scientific advance. These may be summarized as follows:

1. *Gradual relinquishment of the quest for certainty*, with replacement of absolutism by an alternative commitment to conceptual relativism; this trend may be associated with the *maximization of cognitive freedom.*
2. *Continuing accretion of cognitive controls*, providing an increasingly adequate collection of criteria for testing the admissibility of alternative conceptual objectifications; the corresponding trend points toward an *optimization of cognitive control.*
3. *Concrescence of previously distinct theories*, with amalgamation of specialized disciplines and supposedly disparate methodologies resulting from generalization of the fundamental objectives of rational inquiry; this trend may be interpreted as *maximization of scope.*

Maximum freedom, optimal control, and maximum scope may now be viewed as constituting minimally sufficient criteria of *optimal organization*, ensuring, *inter alia:*

(a) long-term viability, based on modifiability of organizational structure and function (cf. maximum freedom);
(b) survival capability in the face of immediate threat, based on effectiveness of adaptive tactical action in the attainment of goals and the reduction of stress (cf. optimal control); and
(c) stable maintenance of these ideal capabilities under continuous expansion of the exploratory domain of interaction, thus preempting the potentially disastrous effects of drastic changes in the environment (cf. maximum scope).

ISBN 0-201-03438-7/0-201-03439-5pbk.

Expanding creativity, rationality, and universality which mark scientific advance are correlates of maximum freedom, optimal control, and maximum scope; and like their correlates, they constitute *simultaneous but antithetical* desiderata. The supreme strategic issue for the cognitive agent will therefore always concern appropriate trade-off among these value measures which cannot be maximized simultaneously by any single course of action. Thus, the cognitive system can never come to equilibrium in terms of its own intrinsic values.

Emerging from this collection of historic trends in science is a continuously improved pattern of cognitive organization, a developing design as awesome as that of evolutionary nature, since it is *of* nature. A new reading of the significance of this evolutionary development of the cognitive modality is immediately within our grasp. Underlying the trends we have noted, we see with some clarity the beginning of the commitment of *Homo sapiens* to a profoundly promising policy: a policy of conscious, purposeful alignment with the process of emergence.

Beyond thinking for the sake of acting, beyond thinking for the sake of thinking, there is a third thinking for the sake of new modalities of thought. This is thought consciously aimed at optimization of the cognitive modality, at realization of successively more flexible, more general, more satisfactory modes of comprehension and action. There is perhaps no more ardent version of purposeful alignment with the process of emergence.

REFERENCES

Marney, Milton, and Anthony, Robert W. (1974). "Modelling Adaptive Systems: Rudiments of Two Normative Theoretic Approaches," paper presented to the Institut de la Vie World Conference "Vers un Plan d'Actions pour l'Humanité," Paris. Conference proceedings (M. Marois, ed.) in press. Amsterdam: North-Holland.

Marney, Milton, and Smith, Nicholas M. (1972). *Foundations of the Prescriptive Sciences*, 2 vols. McLean, Virginia: Research Analysis Corporation.

ISBN 0-201-03438-7/0-201-03439-5pbk.

10

Toward Cultural Symbiosis

Magoroh Maruyama

In the past, the members of a culture did not need to generate cultural goals. In most cases, cultures were either stationary or very slowly changing, and cultural goals were transmitted from the older generation to the younger generation in the process of socialization. Sudden culture change, which did occur from time to time, was merely a matter of transition from one stationary or almost stationary pattern to another stationary or almost stationary pattern.

But we are now entering an era of transition of a different nature. It is a transition from a chain of stationary or quasi-stationary patterns, which the population accepted as given, to a *duration of perpetually transforming patterns* which depend on people's will and choice. It is a transition between *types* of transitions. This can be called a *metatransition.*

In the past, education could aim at transmission of relatively stationary goals and of relatively known means to attain the goals. Education could be considered as information-giving and answer-giving. This type of education will become inadequate for people preparing to enter the period of nonstationary cultures. We must unlearn to expect information on ready-made goals and means. Education will increasingly become a matter of developing an attitude, ability, and skills to transcend the existing cultural goals and means and to challenge our present ways of thinking, logic, science, and epistemology.

1. MUTUAL CAUSALITY AND SOCIAL PROCESS

Aside from this metatransition, our era is also characterized by an epistemological transition. By "epistemology" we mean here the framework and the internal structure of the process of reasoning. The basic epistemology of the American culture was essentially derived from the Greek-European epistemology based on deductive logic, assumption of one-way causal flow, and hierarchical social order mixed with the pecu-

ISBN 0-201-03438-7/0-201-03439-5pbk.

Erich Jantsch and Conrad H. Waddington (eds.), *Evolution and Consciousness: Human Systems in Transition.*

liarly American world view (Mead, 1942, 1946) of unidimensionally rankable universe, competition, conquest, technocentrism, and unicultural assimilation. Under this epistemology, even so-called democracy took the form of majority rule over minorities, that is, domination by quantity.

This epistemology is being challenged by an emergence of other epistemologies. Some of these epistomologies have long existed among the ethnic minority groups in the United States unrecognized by the social majority: for example, the nonhierarchical mutualism of Navajos (Dyk, 1938; Kluckhohn, 1949; Maruyama, 1967) and Eskimos; the philosophy of balance of nature among most of the Native Americans (American Indians); and the world view of mutual complementarity brought in by Chinese and Japanese immigrants. Hippies, rebelling against traditional Americanism, borrowed these epistemologies, imported Zen practice from India and the kibbutz social structure from Israel, and added some contributions of their own, such as psychedelic arts. Ethnic consciousness sprang up in various groups, and the black–white dichotomy has now turned into an impetus toward cultural diversification and political pluralism.

Improved birth control methods led to increased experimentation with new forms of sexual relationship, such as multifamilies and multimarriages. The nuclear family system tends to foster a hierarchical, monocephalic, and uniformistic view of the universe (Maruyama, 1966). It can be expected that multifamilies and multimarriages will enable children to develop more naturally a nonhierarchical, mutualistic, and pluralistic epistemology.

Thus, our society is suddenly faced with not only one but several sets of epistemological discrepancies between established ways of thinking and emerging ways of thinking. Most conspicuous of these discrepancies are (1) competition versus sharing; (2) technocentric transgression versus the harmony of nature; (3) material efficiency versus cultivation of the mind; (4) hierarchism versus nonhierarchical mutualism; (5) the concept of leadership versus interactionism; (6) majority rule versus the elimination of hardship on any one individual; (7) homogenization versus pluralism.

In a heterogeneous society with a wide range of varying cultural niches, an individual can seek out an existing group with which he finds resonance, or several individuals in resonance may form a new group. Individuals—heterogeneous or homogeneous, belonging to different groups, the same group or no group—interact among themselves and influence one another's goals.

Until recently, the concept of mutual causality was a taboo in Western thinking. This was because the structure of events was confused with the structure of the Aristotelian logic which prohibited "cir-

ISBN 0-201-03438-7/0-201-03439-5pbk.

cular argument." But mutual causality exists in many biological, ecological, physical, and social processes, as various branches of science have begun to recognize gradually. A mutual causal network with feedback loops may be hooked up in such a way that it works to amplify change, counteract change, drift randomly, or any combination of these (Maruyama, 1963a).

The development of the study of feedback systems can be roughly divided into two phases. The dividing line, somewhat oversimiplified, is around 1960. In the 1940s and 1950s engineers and biological scientists focused their attention mainly on deviation-counteracting feedback systems (so-called negative feedback systems), which were useful for automatic control of various engineering devices as well as automatic regulation in biological systems. During the same period, however, a small number of thinkers, for example, the Swedish economist Gunnar Myrdal and the cultural psychiatrist Gregory Bateson, began to conceptualize theories of deviation-amplifying mutual causal processes (so-called positive feedback systems). Myrdal published his *American Dilemma* (1943)—a study of economic vicious circles in the American ghettos—during the Second World War, and his theory of the economy of materially poor countries (1957)—a study of economic vicious circles in materially poor countries and in international trade—after the War. He also emphasized the constructive use of deviation-amplifying mutual causal processes. For example, if the economy in a poor country is given an initial kick in the right direction, the small initial change can be amplified to produce an economic development disproportionally large as compared to the size of the initial kick.

But more detailed laboratory-type research on deviation-amplifying mutual causal processes lagged behind the study of deviation-counteracting mutual causal processes, and became formalized around 1960 (Ulam 1962, Braverman and Schrandt, 1966). Deviation-amplifying mutual causal processes can increase differentiation, develop structure, and generate complexity (Maruyama 1963a, 1963b; Waddington 1968-1972). They are found in the interaction of cells during the growth of the embryo into the adult, in the interspecific and intraspecific interaction in the evolutionary process, and so forth. For example, there is a species of moth, and a species of bird which eats the moth. The mutants of the moth who are more camouflaged than the average survive better, and the moth gets increasingly camouflaged generation after generation. On the other hand, the mutants of the bird who are more clever than the average in discovering the moth survive better, and the bird becomes cleverer and cleverer generation after generation.

Development of a city in an agricultural plain may be understood with the same principle. At the beginning, a large plain is entirely

ISBN 0-201-03438-7/0-201-03439-5pbk.

homogeneous as to its potentiality for agriculture. By some chance an ambitious farmer establishes a farm at a spot on it. This is the initial kick. Several farmers follow the example and several farms are established. One of the farmers opens a tool shop. Then this tool shop becomes a meeting place of farmers. A food stand is established next to the tool shop. Gradually a village grows. The village facilitates the marketing of the agricultural products, and more farms flourish around the village. Increased agricultural activity necessitates development of industry in the village, and the village grows into a city.

The growth of the city first increases the internal structuredness of the city itself. Second, it increases the inhomogeneity of the plain by its deviating from the original prevailing condition. Third, the growth of a city at a spot may have an inhibiting effect upon the growth of another city in the vicinity, just as the presence of one swimming pool may discourage an enterpriser from opening another pool right next to it, and just as the presence of large trees (i.e., their shade) inhibits the growth around them of some species of small trees. A city needs a *hintergrund* to support it, and, therefore, cities have to be spaced at intervals. This inhibiting effect further increases the inhomogeneity of the plain.

The crucial consideration is the existence of such mutual causal loops *within* a system, which may be a city, a country, or the entire universe. Such loops can increase structuredness *within* the system, and should be considered apart from the question whether the system is open or closed. It is the *internal* interactions that are relevant here.

Ironically, information theory, which Claude Shannon (1949) formulated soon after the Second World War and which contributed greatly to the development of the study of deviation-counteracting feedback systems, was trapped in a classificational epistemology which sees a structure as consisting of elements which tend to behave independently. In Shannon's theory, the most probable or "natural" state of a system is that in which each element flips its own coin regardless of the other elements. The amount of information which a structure conveys is defined as the degree of departure of the structure from such probable states of nonstructure. The higher the improbability of the structure, the higher the amount of information. In this epistemology, structures tend to decay to more probable nonstructures. All Shannon could do was to combat this decay by means of deviation-counteracting feedback systems. Therefore, in Shannon's formulation, evolution and growth of structures were impossible, or so highly improbable that they had to be attributed to something beyond his theory. The crucial point is that according to Shannon's formulation, interactions *within* a system must be seen either as something counternatural in the sense of being man-

ISBN 0-201-03438-7/0-201-03439-5pbk.

made with a purpose, or as something which has a miraculous or inexplicable origin in the sense of being beyond the natural probability of random process.

On the other hand, evolution, growth, and life have become *causalistically* explicable in the mutualistic logic: deviation-amplifying mutual causal processes can increase differentiation, complexity, and structure; and deviation-counteracting mutual causal processes can maintain them. Since the "discovery" of the mutual causal logic in science ("unscientific" cultures had known it much earlier), it has become scientifically clear that *the basic principle of the biological and social universe is increase of diversification, heterogeneity, and symbiotization. What survives is not the strongest, but the most symbiotic.*

We have been misguided, by the traditional mainstream "scientific" logic of unidirectional causality and by the model of classical physics, into believing that generalizability, universality, homogenization, and competition are not only the rules of the universe but also the *desirable* goals of our society. There are several other ways in which we have been misguided by the logic of unidirectional causality. For example, one of the sacred laws in traditional science stated that similar conditions produce similar results. Consequently, dissimilar results were attributed to dissimilar conditions. Many scientific researches were dictated by this philosophy. For example, when a scientist tried to find out why two identical-twin brothers were different, he looked for a difference in their environment or in the influence on them of other persons. It did not occur to him that neither environment nor other persons might be responsible for the difference. He overlooked the possibility that, for example, the interaction between the two brothers may have acted as deviation-amplifying, or that the two brothers, with identical amplifier circuitry within themselves, amplified some very small difference in their experience. Similarly, when a historian or a geographer tried to find why one of two places with identical conditions had become a city and the other had not, he tried to find some large differences in some unknown variable in the initial conditions. He overlooked the possibility that some almost insignificant event, such as a traveler's getting stranded because of his horse's illness, may have acted as an initial kick, on which a system of deviation-amplification worked. In such cases, it is more meaningful to study the circuitry or deviation-amplification, and it would be a waste of time to look for nonexistent large differences in the initial condition. I have applied the methodology of the mutual causal model in my analysis of the Danish culture (Maruyama, 1961a).

The law of causality is now revised to state that similar conditions may produce dissimilar results if there is a deviation-amplifying

ISBN 0-201-03438-7/0-201-03439-5pbk.

mutual causal network. It is important to note that this revision is possible *without* the introduction of indeterminism and probabilism. But if we combine this revision with indeterminism, we obtain the following: A small initial deviation, which is within the range of high probability, may develop into a large deviation of low probability (or more precisely into a large deviation which is very improbable within the framework of probabilistic unidirectional causality).

Many decades ago, the theory of relativity had already challenged the notion of *substance*, and quantum mechanics had questioned the principle of *identity*. Mutual causal logic can challenge the notion of substance and the principle of identity from another angle, and without the help of the theory of relativity or quantum mechanics. It shows precisely how differentiation, growth, and increase of complexity can take place; how heterogeneity can arise out of seeming homogeneity; and how new structures create themselves without a predesigned blueprint. This is a considerable challenge to the notions of permanence, homogeneity, and universal validity, which are some of the consequences of the notions of substance and of identity.

Change-amplifying mutual interaction is a useful concept in our discussion of goal generation. One of the properties of the deviation-amplifying mutual causal system is that a small initial kick may result in a large change. This property has several consequences. One is that two identical systems with almost identical initial conditions may produce very different results, depending on a small difference in the direction of the initial kick or on a minute difference in the initial condition, as discussed above. Another consequence is that a relatively small "investment" of the initial kick will produce a disproportionally large "return." The third consequence is that once an amplification has started, it may be very difficult to stop or reverse the process. The fourth consequence is that when the initial kick is left to inconspicuous random fluctuations, the produced result takes an appearance of unpredictability because of the invisibility of the initial kick. Much of the unpredictability of history can be attributed to deviation-amplifying mutual causal processes rather than to some "unknown" variables.

Goal-generating as a mutual interaction process between individuals, between groups, or between levels of hierarchy possesses these characteristics. These characteristics can be utilized to advantage, but may also cause dangers. For example, an administrative policy of suppressing dissenters tends to widen the gap between the dissenters and the administration instead of reducing it. The people who support the administrative policy reinforce one another's conservatism among themselves, aided by administrative sanction and its perception of the dissenters. This may lead to an impasse and bloodshed.

ISBN 0-201-03438-7/0-201-03439-5pbk.

2. SYMBIOSIS

There are several ways the competent groups can interact in a heterogeneous society. Groups may coexist with no or very little interaction between them. This is separatism. Or groups may benefit from one another. This is symbiosis. Or the relation may be such that what one group gains is what another group loses. This is parasitism. Or one group may harm another group. This is antibiosis. Or many groups may harm one another. This is mutual antibiosis.

These relations can be classified into zero-sum relations and nonzero-sum relations. Zero-sum relations are those in which the total gain and total loss cancel out. Nonzero-sum relations are those which are not zero-sum relations. Table 10.1 summarizes these relations.

TABLE 10.1. Group Relations in a Heterogeneous Society

Relation	*Gain*	*Loss*	*Remarks*
Separatism	No group	No group	No interaction
Symbiosis	All groups	No group	Nonzero sum
Parasitism	Some groups	Other groups	Zero sum
Antibiosis	No group	Some groups	Nonzero sum
Mutual antibiosis	No group	All groups	Nonzero sum

The ideal relationship, of course, is symbiosis. There are two ways to conceptualize symbiosis. One is organismic (Pribram, 1949), and the other is mutualistic.

The organismic view assumes that there is a whole to which the parts are subordinated. The parts cooperate with one another in order to fulfill the task of the whole. There is a hierarchical, causal, and teleological priority of the whole over the parts. The whole is the cause of the parts. The whole is more important than the parts.

The mutualistic view, on the other hand, is that there are only parts, and parts create a system of interaction. There is no "whole" prior to the parts.

The strategy of planning in a heterogeneous society varies, depending upon whether the organismic view or the mutualistic view is used. In organismic planning a hierarchical planning center is set up, and the overall policy proceeds from the top down. The component groups conform to the outline issued by the center. In mutualistic planning, the ideas originate from the component group and are pooled for mutual adjustment. Solutions are explored in various dimensions until one is found which satisfies all the groups or at least does not impose hardship on any group. No majority vote is taken to be imposed on minorities.

ISBN 0-201-03438-7/0-201-03439-5pbk.

In a heterogenistic society, symbiotic combinations of diverse individuals or groupings must be worked out. Suppose Individual A has Goal A, Individual B has Goal B, Individual C has Goal C, and so on. Individual A has several alternative ways a_1, a_2, a_3, $\cdots$ to implement his Goal A. Individual B has several alternative ways b_1, b_2, b_3, $\cdots$ to implement his Goal B, and similarly for other individuals. Some combinations $a_i b_j c_k \cdots$ of these alternative ways may be symbiotic, whereas other combinations may not be. Individuals for whom a symbiotic combination can be found among the alternative ways of their respective goal implementation can be hooked up in a network of symbiosis. Individuals who cannot be combined symbiotically can be hooked up in separate networks. *Each* of these networks is *heterogeneous.*

On the other hand, individuals who share a *similar* goal may be combined in a grouping. A grouping is therefore homogenous in goal, and can be treated like an individual in the network analysis. A grouping does not have to be geographic or ethnic. It can be occupationally, recreationally, or otherwise oriented. Groupings for which a symbiotic combination can be found among the alternative ways of their respective goal implementation can be hooked up in a network of symbiosis. One of the tasks of human futuristics can be to help develop mutualistic symbiosis in a heterogeneous society.

There are two ways in which heterogenization may proceed: *localization* and *interweaving.* In localization, heterogeneity between localities increases, while each locality may remain or become homogeneous. In interweaving, heterogeneity in each locality increases, while the differences between localities decrease. At present, localization is more conspicuous than interweaving. In the next stage of our social heterogenization, interweaving may increase. We need to develop principles of heterogenization and symbiotization of heterogeneity for both localization and interweaving.

There is already a great deal of data from ecology and biology regarding symbiosis and heterogeneity. But so far in ecology and biology we have been studying the symbiotic relations which are *already established* between diverse species. Nature arrives at symbiotic relations at the cost of the extinction of those species which are less symbiotic. Furthermore, nature does not look for possible symbiotic relations between species which are not yet interacting.

In our changing society, new life patterns and new social patterns are emerging, and individuals or groups who have so far been symbiotic may become parasitic or antibiotic. On the other hand, those who have been parasitic or antibiotic might possibly become symbiotic. Moreover, individuals or groups who have not yet been interacting are

ISBN 0-201-03438-7/0-201-03439-5pbk.

coming into contact. In these situations, we do not wait for nonsymbiotic individuals or groups to die off. We must prevent parasitic or antibiotic combinations from occurring. If they can be recombined into symbiotic relations, we must so recombine them. If such recombinations are impossible, the parasitic and antibiotic individuals or groups must be separated into different networks. We also have to look for possible symbiotic combinations among those who are not yet interacting.

3. THE NEED FOR TRANS-EPISTEMOLOGICAL PROCESS

Our social thinking differs from our ecological and biological thinking in another aspect. In most cases ecological and biological thinking proceeds on two assumptions: (a) that animal species and plant species do not think about alternatives; they live the way they live; and (b) that the ecological relationship which is established among them should be maintained as much as possible. A change in any of its parts may affect the whole ecosystem, and may have disastrous or irreversible consequences.

On the other hand, our social thinking proceeds on two different assumptions: (a) that each culture, each social group, or each individual has its own goal, and there are several *alternative ways* to attain this goal; and (b) that current international relations, intercultural relations, intergroup relations, power structure, and so on are by no means satisfactory. We must find more satisfactory ways to rearrange these relations and structures.

The existing models of social planning may have shortcomings: (a) they are culture bound and paradigm bound and cannot be applied to different cultures; (b) they are homogenistic, and cannot even deal with the heterogeneity in our own culture; (c) consequently, they do not even conceive the problem of finding possible symbiotic combinations among heterogeneous elements; and (d) not many of them include the consideration of possibilities of alternatives.

Therefore, we need a *trans-epistemological process*: we need to develop the citizens' and planners' ability to grow out of the traditional epistemology and explore, discover, invent, and reality-test new epistemologies. We also need to design built-in processes within our social system to transcend its own operational logic.

During the past decade we have seen the emergence and increasing acceptance of a new type of logic in many segments of American society. The hippie movement of affluent youths, the ethnic movements of

ISBN 0-201-03438-7/0-201-03439-5pbk.

oppressed minority groups, and the ecology movement, which finds supporters among both conservatives and liberals in the middle class, have all converged to a logic opposed to the traditional mainstream American logic (see Table 10.2).

TABLE 10.2

Traditional mainstream logic	*Emerging logic*
Unidirectional	Mutualistic
Uniformistic	Heterogenistic
Competitive	Symbiotic
Hierarchical	Interactionist
Quantitative	Qualitative
Classificational	Relational
Atomistic	Contextual

These three segments of our society—affluent youths, oppressed minority groups, and the middle class—did not arrive at this new logic for the same reason, nor did they necessarily perceive one another to be converging on the same logic. The Black ghetto residents who were *materially*, though not culturally, deprived, and who had to strive for material improvement of their life, tended to regard hippies as silly rich kids trying to go backward materially, and to consider pollution as a rich people's problem. Conversely, the hippies, though antiestablishment, did not necessarily have much insight into the nature of the type of oppression imposed upon the Black ghetto residents.

Therefore, the convergence of these three segments of society toward a common logic is a result of mutually *independent* recognition, on the part of the three segments, of the inadequacy of the traditional mainstream logic. As Gerlach (1976) pointed out, even within each of the segments the movements *began from small, independent groups*, which only *later* became more and more coordinated. This recognition of the inadequacy of the traditional mainstream logic from such widespread independent sources is an indication that the total structure, not component structures, of our society has become obsolete.

Several forms of mutualistic logic have existed for a few thousand years in Asian and African cultures as well as in many of the American Indian cultures. The principles of symbiotization of dissimilar elements have also existed in traditional Japanese garden design and flower arrangement, though the Japanese imported a Vitruvius-like principle of city planning from China for the design of the ancient capitals at Nara and Kyoto.

One of the factors contributing to the persistence of one-way logic in Western cultures seems to be the nuclear family system, which en-

ISBN 0-201-03438-7/0-201-03439-5pbk.

courages *monopolarization* in personality development (Maruyama, 1966). Monopolarization is a psychological need to believe that there is *one* universal truth, and to seek out, depend on, find security in, and hang onto *one* authority, *one* theory, uniformity, homogeneity, standardization, and so on. In cultures in which the children have close relations with many adults—whether they are in extended families, communal child rearing, a system of frequent child exchange, or a system in which a person's obligations to many specific individuals are emphasized—there seems to be less tendency toward monopolarization. It is not a coincidence that in the current epistemological revolution toward mutualistic logic in our society, many nonnuclear forms of family are experimented with.

The mainstream logic in Western cultures has been Aristotelian logic, even though countermovements have occurred at different times in history. This logic is homogenistic, and has a *monocular vision.* Vision with one eye has no depth perception. On the other hand, vision with two eyes has depth perception, not because one of the eyes can see the back side of the object, but because the *differential* between the two images enables the brain to compute the third, not directly visible, dimension. Binocular vision allows us to see space three dimensionally.

Suppose we live in a four-dimensional space. Perhaps we then need three eyes in order to have a four-dimensional perception. In an N-dimensional space, we might need $(N-1)$ eyes. This is the metaphorical rationale of *polyocular vision.*

Many cultures, such as the Mandenka culture in West Africa (Camara, 1975), have a heterogenistic logic. The universe in these cultures is based on heterogenistic principles, and there are as many different ways to see the same situation as there are individuals. There is no assumption of the existence of one "objective" truth or one "correct" point of view. Instead, people understand the situation cross-subjectively and polyocularly.

The exact number of eyes needed for N-dimensional perception is not important because it is only a metaphor. What is nonmetaphorically important is the consideration that the *differentials* between different views enable us to compute the dimensions which are not directly observable.

Let me emphasize that Westerners tend to talk about "perspective" in *quantitative* ways. There are three different ways in which the Westerners use the word perspective, and they are all quantitative. The first is to use the word in the sense of the scale of a map or the magnification factors of a microscope or a telescope. The second is to use it in the sense of different "parts," as when several blind people touch dif-

ISBN 0-201-03438-7/0-201-03439-5pbk.

ferent parts of en elephant. The third is to use it in the sense of different "angles" from which the same object is viewed. In all three cases, the "information" from different "perspectives" can be *added up* to produce a more "comprehensive" picture.

On the other hand, the perspective in polyocular vision more qualitative. It is more like different color overlays of the same picture. But this is still an inadequate metaphor because color overlays can be additive, whereas what is important in polyocular vision are the *differentials.* A closer metaphor would be the stereophonic record player; the *difference* in sound quality between two speakers enables you to hear the music three dimensionally.

Monocular vision is an impoverishment as compared to binocular vision or polyocular vision. Furthermore, if you eliminate the parts of the images on which the two eyes "disagree," what is left would be much less than the single image obtained in monocular vision. This is why insistence on "objective" agreement on *one* view impoverishes our perception and reduces us to even much less than monocular vision. Similarly, monopolarization is an impoverishment of human life.

4. STEPS TOWARD A NEW LOGIC

As mentioned, we are in a transitional period from the 2500-year-old traditional Western logic to a new logic. Such a transition in logic may be called an epistemological transition. It is more than a transition from one paradigm to another. For monopolarized persons, it is very difficult to undergo paradigmatic or epistemological transitions. For them, being confronted with other ways of thinking is a traumatic experience. If they realize that there are other ways of thinking, their "truth" is called into question and they would have to feel as if the whole universe were collapsing. Many of them react to this trauma by reinforcing their own belief and becoming extremely defensive.

The first step is to reduce this defensiveness and accept the *trauma of demonopolarization* as a positive experience without defensiveness. Painful as it may be to some persons, this first step must be taken.

The second step is trans-spection (Maruyama, 1969): to get into the *head* (not shoes) of another person. Instead of disagreeing, you try to think exactly like the other person. Care must be taken not to reduce the other person's logic to your own logic. Beginners often fall into the delusion of being able to think in the other person's logic, whereas in fact all they are doing is reducing the other's logic to their own logic.

ISBN 0-201-03438-7/0-201-03439-5pbk.

Many methods of feedback and checks must be made with the other person until trans-spection is attained (Maruyama, 1961b, 1962, 1963c).

The third step is to let the other person trans-spect into your own logic. This is not as easy as it sounds. If you have never trans-spected into other logics, you would not even know how your logic is different from other logics. This is why the third step should not begin until the second step is reasonably completed.

We also need to create new logics and new paradigms which do not yet exist. Recently the need for this process has been repeatedly emphasized (Harman, 1971; Boulding, 1976; Maruyama, 1976). We do not yet have a methodology for the creation of new logics and paradigms. Perhaps a methodology, once established, would only limit the types of logic and paradigms it can create.

One useful consideration is the principle of creativity by interaction. It is a fallacy to equate creativity with capriciousness. A symphony is the opposite of random noise. Creativity involves the development of patterns, differentiation, and structure. This is possible by means of differentiation-amplifying mutual causal processes, as we have seen. The amount of Shannonian information can *increase* in such processes. Interaction can create *new* patterns, not just new combinations of old patterns.

Mead (1964) analyzed how informal networks among scientists produce innovations. Wright (1931) studied how the speed of evolution is related to the ratio between the mutation rate and the size of an interactive population. When the mutation rate is low and the population size is large, interbreeding between mutants and "normal" individuals, or between mutants of opposite directions, tends to cancel out the mutations, and the speed of evolution is very slow. On the other hand, when the mutation rate is very high and the population size is small, there is much inbreeding, which tends to amplify the mutations rapidly, and evolution may take place so fast that the newly evolved species has time neither to adapt to the new environment, to seek an appropriate new environment, nor to work out new types of relationships between individuals, and the whole species may become extinct. When the mutation rate is moderate and the population size is neither too large nor too small, the mixture of stabilization and change takes place, and the phenomenon of random drift occurs: a change may be amplified for a while, stabilization takes place, another change in a new direction takes place, and so on. A faster rate of evolution occurs when the total population is subdivided into semi-isolated colonies, connected by occasional interbreeding.

If a number of similar-minded innovators get together and seclude themselves from innovators of different minds, as well as from the rest

ISBN 0-201-03438-7/0-201-03439-5pbk.

of the society, the situation is similar to that of a high mutation rate within a small population. The peculiarities within the group may be amplified so fast that there is no time to work out the details to fit together within the new structure, not to mention the adjustment of the group to the rest of society.

On the other hand, if small networks of innovators develop independently in many different cultures or different social contexts, with some exchange of ideas between networks, while avoiding excessive mutual imitation, the speed of intellectual evolution can be greater than otherwise. Such semi-independent subpopulations have another advantage: they allow for multibranch evolution and heterogeneity, and this increases the probability of the survival of the species as a whole in case of unexpected disasters.

Since intellectual exchange does not require the physical contact that is necessary in biological reproduction, the subpopulations in intellectual communication do not have to be defined geographically. An intellectual network may include individuals from many different countries and cultures, and exclude many individuals within the same geographic location.

Whether localized or interwoven, heterogeneity and interaction are the key to the creation of new logics and paradigms.

5. CONCLUSION

The basic principle of biological and social, and even some physical, processes is increased heterogeneity, symbiotization, and evolution. This is possible by means of differentiation-amplifying mutual causal processes. In the past, there have been many cosmologies based on different logical models. The Parmenidean universe of *static permanence* was recaptured, in a different form, by the Newtonian universe of *unchanging movements* of astronomical bodies. The Heraclitean universe of *incessant change* did not specify the direction of change. The medieval theologians formulated a *hierarchical* universe with a creator and prime mover. The second law of thermodynamics, and the later theory of information by Shannon, were based on the assumption that the *most probable* state of the universe is that of a random distribution of events, each of which tends to behave with its own probability independently. Their universe was an ever-decaying one. This universe was based on the *logic of basically independent events.* On the other hand, cybernetics brought forth the *logical model of mutual causal loops* within a system. Its first phase as a study of negative feedback systems led to a view of deviation-counteracting, decay-combating, equilibrium-

ISBN 0-201-03438-7/0-201-03439-5pbk.

seeking, or converging universe. In its second phase, deviation-amplifying mutual causal systems with internal positive feedback loops began to be studied intensively. Such systems can lead to either a runaway situation or evolution. Often the runaway processes have been emphasized.

The purpose of this chapter has been to bring back in focus the evolving, self-generating, and self-organizing universe based on differentiation-amplifying mutual causal logical models. We need to recognize and make use of the principles of heterogenization and symbiotization for our intercultural ecology and cultural evolution.

REFERENCES

Boulding, E. (1976). "Futuristics and the Imaging Capacity of the West," in *Cultures of the Future* (M. Maruyama and A. Harkins, eds.). The Hague: Mouton.

Braverman, M. H., and Schrandt, R. G. (1966). "Colony Development of a Polymorphic Hydroid as a Problem in Pattern Formation," in *Cnidaria and Their Evolution (Symp. Zool. Soc. London,* **16***).*

Camara, S. (1975). "The Concept of Heterogeneity and Change among the Mandenka," *Technological Forecasting and Social Change*, **7**, 273-284.

Dyk. W. (1938). *Son of Old Man Hat.* New York: Harcourt.

Gerlach, L. (1976). "Fumbling Freely into the Future," in *Cultures of the Future* (M. Maruyama and A. Harkins, eds.). The Hague: Mouton.

Harman, W. W. (1971). "Planning amid Forces for Institutional Change," paper presented at Amer. Soc. Public Administration.

Kluckhohn, C. (1949). "The Philosophy of Navaho Indians," in *Ideological Differences and World Order* (F. S. C. Northrop, ed.). New Haven: Yale Univ. Press.

Maruyama, M. (1961a). "The Multilateral Mutual Simultaneous Causal Relationships among the Modes of Communication, Sociometric Pattern and Intellectual Orientation in the Danish Culture," *Phylon*, **22**, 41-58.

Maruyama, M. (1961b). "Communicational Epistemology," *Brit. J. Philos. Sci.*, **11**, 319-327; **12**, 52-62; **12**, 117-131.

Maruyama, M. (1962). "Awareness and Unawareness of Misunderstandings," *Methodos*, **13**, 255-275.

Maruyama, M. (1963a). "The Second Cybernetics: Deviation-Amplifying Mutual Causal Processes," *Amer. Scientist*, **51**, 164-179; **51**, 250-256.

Maruyama, M. (1963b). "Generating Complex Patterns by Means of Simple Rules of Interaction," *Methodos*, **14**, 17-26.

Maruyama, M. (1963c). "Basic Elements in Misunderstandings," *Dialectica*, **17**, 78-92; **17**, 99-110.

ISBN 0-201-03438-7/0-201-03439-5pbk.

Maruyama, M. (1966). "Monopolarization, Family and Individuality," *Psychiatric Quarterly*, **40**, 133–149.

Maruyama, M. (1967). "The Navajo Philosophy: An Esthetic Ethic of Mutuality," *Mental Hygiene*, 51, 242–249.

Maruyama, M. (1969). "Epistemology of Social Science Research: Explorations in Inculture Researchers," *Dialectica*, 23, 229–280.

Maruyama, M. (1976). "Toward Human Futuristics," in *Cultures of the Future* (M. Maruyama and A. Harkins, eds.). The Hague: Mouton.

Mead, M. (1942). *And Keep Your Powder Dry: an Anthropologist Looks at America.* New York: Morrow.

Mead, M. (1946). "An Application of Anthropological Techniques to Cross-national Communication," *Trans. N. Y. Acad. Sci. Ser. 2*, **9**, 133–152.

Mead, M. (1964). *Continuities in Cultural Evolution.* New Haven, Conn.: Yale Univ. Press.

Myrdal, G. (1943). *American Dilemma.* New York: Harper and Row.

Myrdal, G. (1957). *Economic Theory and Underdeveloped Regions.* London: Duckworth.

Pribram, K. (1949). *Conflicting Patterns of Thought.* Washington, D.C.: Public Affairs Press.

Shannon, C. and Weaver, W. (1949). *Mathematical Theory of Communication.* Urbana, Ill.: Univ. of Illinois Press.

Ulam, S. (1962). "On Some Mathematical Problems Connected with Patterns of Growth Figures," *Proc. Symp. Appl. Math.*, **14**, 215–224.

Waddington, C. H. (ed.) (1968-1972). *Towards a Theoretical Biology.* 4 vols. Edinburgh: Univ. of Edinburgh Press; Chicago: Aldine (Vols. 1, 2); New York: Halsted Press (Vol. 3).

Wright, S. (1931). "Evolution in Mendelian Population," *Genetics*, **16**, 97–159.

ISBN 0-201-03438-7/0-201-03439-5pbk.

11

Human Consciousness in Transformation

O. W. Markley

1. THE IMPORTANCE OF GUIDING IMAGES

A variety of writing throughout history indicate that the underlying images held by a culture or a person have an enormous influence on the fate of the holder (Markley et al., 1974). We use the term "image of man" (or image of man-in-the-universe) to refer to the set of assumptions and fundamental premises held about the human being's origin, nature, abilities, characteristics, relationships with others, and place in the universe. No one knows the total potentiality of humankind, and our awareness of human "nature" (including the nature of human systems) is selective, shaped by our explicit and implicit images.

In a provocative book, *The Image of the Future*, the Dutch sociologist Fred Polak (1973) noted that when the dominant images of a culture are anticipatory, they "lead" social development and provide direction for social change. They have, as it were, a "magnetic pull" toward the future. By their attractiveness and legitimacy they reinforce each movement that takes the society toward them, and thus they influence the social decisions that will bring them to realization. As a culture moves toward the achievement of goals inherent in its dominant images, the congruence increases between the images and the development of the culture itself—the implications of the images are explored, progress is made, and needs are more fully satisfied (see Figure 11.1).

If the "progress" of the human system outstrips that of its traditional images, however, its policies and behavior (which are based on the old dominant images) become increasingly faulty—even counterproductive—precipitating a period of frustration, cultural disruption, or social crisis. The stage is then set for basic changes in the underlying images and the organization of the system. Various indicators suggest that our culture may now be nearing, if not at, such a stage (Markley et al., 1974).

This, of course, assumes that it is reasonable to view history as a series of episodic paradigms, eras or epochs—each having a relatively

ISBN 0-201-03438-7/0-201-03439-5pbk.

Erich Jantsch and Conrad H. Waddington (eds.), *Evolution and Consciousness: Human Systems in Transition.*

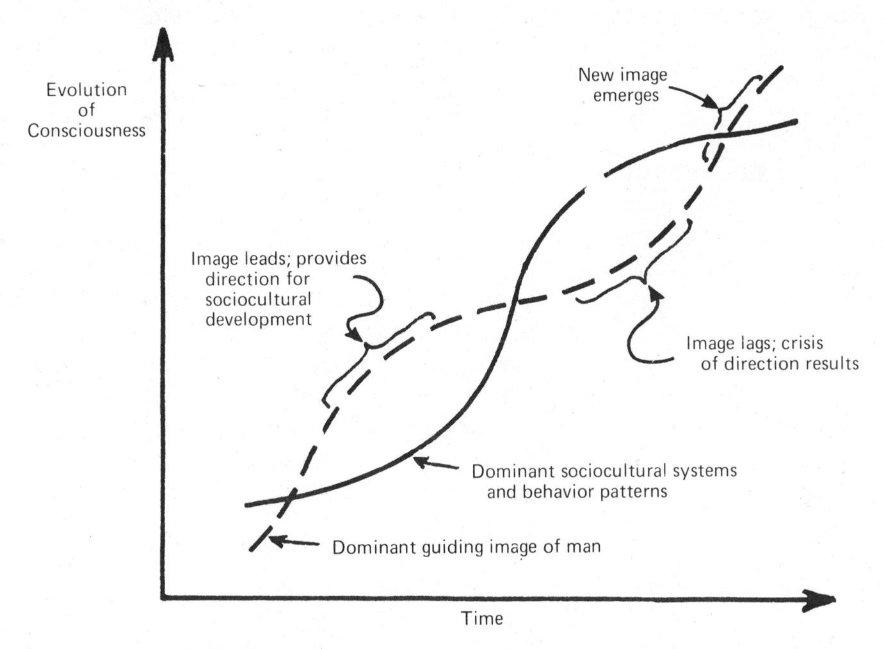

FIGURE 11.1. Hypothesized phase relationship between dominant images and sociocultural development (from Markley et al., 1974, pp. 6, 59).

distinct character—rather than as a continuing series of incremental adjustments and adaptations to emerging conditions. Although a number of arguments can be made to support either view (and to some extent both views are valid), a substantial amount of evidence exists that successful responses to crisis tend to be *nonincremental* in character, and lead to restructured images and modes of system organization.

The focus of this chapter is conceptual exploration of this evolutionary change process and what it implies for the study and application of human systems as an emerging discipline.

2. THE CYCLE OF TRANSFORMATION

When a human system of any scope is faced with a group of problems that are not possible to resolve by using customary assumptions and procedures, then new ways become necessary. If new ways are not discovered and successfully applied, the system becomes unhealthy and may fail to survive at all. To illustrate the conceptual similarity of the processes through which several levels of human system functioning are

ISBN 0-201-03438-7/0-201-03439-5pbk.

transformed in response to crisis, we shall consider five examples—cultural revitalization, revolutions in science, heroic mythology, psychotherapy, and creativity.

Cultural Revitalization

In a cross-cultural study, the anthropologist Anthony Wallace (1956) examined crisis-motivated cultural changes that have occurred at various times and places. He then derived a series of idealized stages through which many such transformations—if successful—have passed. Especially relevant for our purposes are Wallace's findings on the process through which images of the role of self and society have often changed in other societies in response to crisis. He discovered that, unlike classic culture change, the process of *revitalization requires explicit intent by members of the society and often takes place within one generation:*

> The structure of the revitalization process, in cases where the full course is run, consists of somewhat overlapping stages: (1) Steady State; (2) Period of Individual Stress; (3) Period of Cultural Distortion; (4) Period of Revitalization (in which occur the functions of mazeway reformulation, and routinization); and finally (5) new Steady State [p. 264].

The key element in the process of transformation is what Wallace terms the "mazeway," which is almost synonymous with the more familiar notion of "image of man-in-the-universe:"

> It is . . . functionally necessary for every person in society to maintain a mental image of the society and its culture, as well as of his own body and its behavioral regularities, in order to act in ways which reduce stress at all levels of the system. The person does, in fact, maintain such an image. This mental image I have called the "mazeway," since as a model of the cell-body-personality-nature-culture-society system or field, organized by the individual's own experience, it includes perceptions of both the maze of physical objects in the environment (internal and external, human and nonhuman) and also of the ways in which this maze can be manipulated by the self and others in order to minimize stress. The mazeway is nature, society, culture, personality, and body image as seen by one person . . . . Changing the mazeway involves changing the total *Gestalt* of his image of self, society, and culture, of nature and body, and of ways of action. It may also be necessary to make changes in the "real" system in order to bring mazeway and

ISBN 0-201-03438-7/0-201-03439-5pbk.

> "real" system into congruence. *The effort to work a change in mazeway and "real" system together so as to permit more effective stress reduction is the effort at revitalization*; and the collaboration of a number of persons in such an effort is called a revitalization movement [pp. 266ff.; emphasis added].

Whether the revitalization movement is religious or secular, the reformulation

> . . . seems to depend on a restructuring of elements and subsystems which have already attained currency in the society and may even be in use and which are known to the person who is to become the prophet or leader. The occasion of their combination in a form which constitutes an internally consistent structure, and of their acceptance by the prophet as a guide to action, is abrupt and dramatic, *usually occurring as a moment of insight, a brief period of realization of relationships and opportunities.* These moments are often called inspiration or revelation. The reformulation also seems normally to occur in its initial form in the mind of a single person rather than to grow directly out of group deliberation [p. 270; emphasis added].

After mazeway reformulation comes a series of steps—communication, organization, adaptation, cultural transformation, and routinization—through which the idealism of the original vision is applied and modified in repsonse to cultural feedback. This idealism tends to be preserved only in those areas where the movement maintains responsibility for the preservation of doctrine and performance of ritual. In other words, the movement becomes a "church," whether religious or secular.

Conceptual Revolutions in Science

Studying the history of science, Thomas Kuhn (1962) recognized a similar pattern. His use of the term "knowledge paradigm" is very similar to Wallace's use of the term "mazeway." Knowledge paradigm is used to denote

> . . . the collection of ideas within the confines of which scientific inquiry takes place, the assumed definition of what are legitimate problems and methods, the accepted practice and point of view with which the student prepares for membership in the scientific community, the criteria for choosing problems to attack, the rules and standards of scientific practice [Kuhn, 1970, p. 11].

ISBN 0-201-03438-7/0-201-03439-5pbk.

Such a knowledge paradigm bears the same relation to the laws and rules in a field of scientific inquiry that myths and rituals bear to the norms of prescientific society. It has a well-understood set of exemplars and precedents that define a field of inquiry, determine the rules that govern the formulation of new problems, and specify acceptable forms of solutions. Thus, the paradigm can exist only if there is a shared commitment to certain beliefs, such as that the molecules of a gas behave like tiny elastic billiard balls, or that certain kinds of procedures should be used for experimentation, or that some topics are appropriate for scientific investigation and others not. Its communicants must also agree on the meaning of symbolic representation, as in mathematics. Finally, its communicants must share relevant values, such as the importance of making predictive rather than nonpredictive explanations, the appropriateness of responding to social concerns during problem formulation, and the degree of simplicity demanded in theories.

The excitement generated by Kuhn's work rests not so much with his formulation of the knowledge paradigm, however, as with his portrayal of the dynamics with which such paradigms are created and replaced. In Kuhn's view, *normal* science does not aim at novelty; rather, it attempts to actualize the promise offered by the existing paradigm. But it results almost invariably in the exposure of anomalies between expectations based on the paradigm and those based on fact. As such anomalies grow more numerous, we see the recurring emergence of crises and the development of new paradigms that embrace both the old paradigms and the anomalous data with which the old could not deal adequately. Kuhn noted that this transformational process typically passes through four stages: pre-paradigm research, normal science, crisis, and revolution, leading again to normal science within a changed paradigm.

Mythology

As various scholars have noted (e.g., Boisen, 1936; Erikson, 1958), often those individuals who bring the new reconceptualizations to society have had personal problems that were similar in form or were significantly related to those of the larger society. In resolving their own problems, they presented viable resolutions to the problems of their culture. This characterization of the hero is in fact so common throughout the transformation myths of different times and places that Joseph Campbell (1956) used the term "the monomyth" to describe it:

> The standard path of the mythological adventure of the hero is a magnification of the formula represented in the rites of passage: *separation*

ISBN 0-201-03438-7/0-201-03439-5pbk.

> —*initiation—return*: which might be named the nuclear units of the monomyth . . . [1956, p. 30].
> The composite hero of the monomyth . . . and/or the world in which he finds himself suffers from a symbolical deficiency. In fairy tales this might be as slight as the lack of a certain golden ring, whereas in apocalyptic vision the physical and spiritual life of the whole earth can be represented as fallen, or on the point of falling, into ruin. Typically the hero of the fairy tale achieves a domestic microcosmic triumph, and the hero of myth a world-historical, macrocosmic triumph. Whereas the former—the youngest or despised child who becomes the master of extraordinary powers—prevails over his personal oppressors, the latter brings back from his adventure the means for the regeneration of his society as a whole [1956, p. 38].

The basic pattern is clear:

> Whether the hero be ridiculous or sublime, Greek or barbarian, Gentile or Jew, his journey varies little in essential plan. Popular tales represent the heroic action as physical; the higher religions show the deed to be moral; nevertheless, there will be found astonishingly little variation in the morphology of the adventure, the character roles involved, the victories gained [1956, pp. 37ff.].

Psychotherapy

Just as the mythological hero often suffers from a defect that spurs him to action, so many of the "great men" of history have suffered from defects—for example, often having been denied in childhood the benefits of carefree, well-adjusted homes (Goertzel and Goertzel, 1962). Such persons do not typically adjust by conforming to personal and social realities that to them seem filled with anomalies. Rather, they attempt to resolve the dissonant elements of their lives in creative ways, which is the central goal of psychotherapy.

Although the literature of psychotherapy is so varied that no generalization can fit all situations, a pattern does emerge from writers who attempt to describe the process of crisis-motivated personal transformation. From the writings of Boisen (1936), Martin (1955), Sullivan (1953), Fingarette (1963), and Kantor and Herron (1966) can be derived the following stages that seem to typify this process (see Markley et al., 1974, pp. 190ff.):

1. Adequate mastery of one's life: reliance on defense mechanisms (e.g., denial, repression, sublimation).
2. Inadequate mastery of one's life: anxiety and disintegration.
3. Looking for causes: blame and guilt.

ISBN 0-201-03438-7/0-201-03439-5pbk.

4. Finding causes: acceptance of responsibility.
5. Looking for new solutions: openness to seeing things anew in both the inner and the outer world.
6. Finding new solutions: insights that reformulate one's existential conceptions and reintegrate the personality.
7. Applying new solutions: learning new modes of behavior that test and apply the new perspective with increased mastery of one's life.
8. New level of adequacy: open-ended growth and learning as normal behavior.

Creativity

It should not be too surprising that the creative process sometimes follows the same general stages, because basic creativity must somehow be at the core of transformational processes. Here the key terms are *preparation*, *incubation*, *illumination*, and *verification* (G. Wallace, 1926).

After conventional approaches are tested and found wanting (preparation), the next step often necessitates making what P. W. Martin (1955) has termed "the experiment in depth." Here one deliberately sets aside the assumptions that are conventionally made about reality, and engages in techniques or activities that open up one's self to more primal and direct perceptions of reality. Such perceptions are less strongly filtered by convention, and often result from deliberate or accidental use of altered states of consciousness. Because these sources of creativity are not yet widely understood, access to them is for most persons a rather random and uncontrolled process. Hence the term "incubation," which suggests the cessation of deliberate attempts to *force* insight. Two quotations describe the process:

> Cease striving; then there will be self-transformation.
>
> Chuang-Tzu, *Book XI*

> Whosoever shall seek to gain his life shall lose it;
> but whosoever shall lose his life shall preserve it.
>
> Luke *17:33*

Homologous Description

Before considering the remaining stages of the basic creative process, the reader may find it useful to pause and to reflect on what has been said thus far. By now a common pattern should be apparent. Table 11.1 summarizes the idealized stages of each of the varied processes through which human systems become transformed in response to crisis. As can be clearly seen, the processes are, to a striking degree, homologous.

ISBN 0-201-03438-7/0-201-03439-5pbk.

ISBN 0-201-03438-7/0-201-03439-5pbk.

TABLE 11.1 The Transformation Cycle in Myth, Culture, Science, Psychotherapy, and General Creativity

Monomyth (J. Campbell)	*Cultural Revitalization (A. Wallace)*	*Scientific Revolution (T. Kuhn)*	*Psychotherapy (O. Markley)*	*General Creativity (G. Wallace)*
		1. Pre-paradigm research		
	1. Steady state	2. Normal science	1. Normal defense mechanisms	
	2. Period of individual stress	3. Growth of anomalies	2. Anxiety and disintegration	
			3. Blame and guilt	
1. Separation				1. Preparation
	3. Period of cultural distortion	4. Crisis	4. Acceptance of responsibility	
			5. Looking for new solutions	2. Incubation
2. Initiation	4. Period of revitalization Reformulation Communication Organization	5. Revolution	6. Insight/reformulation/reintegration	3. Illumination
3. Return	Adaptation Cultural transformation Routinization		7. Testing and application	4. Verification
	5. New steady state	6. Normal science in new paradigm	8. Open-ended change and growth	

The moment of insight (illumination), as we observed in connection with the cultural revitalization movements and creation of scientific paradigms, occurs with vivid clarity and suddenness, is abrupt and dramatic, "a brief period of realization of relationships" (A. Wallace, 1956, p. 270) that "inundates a previously obscure puzzle, enabling its components to be seen in a new way for the first time" (Kuhn, 1970, pp. 122ff.). Thus, the moment of sudden insight seems to be an element common to radical discovery and transformation—both mythic and scientific. We might well apply to this type of reconceptualization the Greek word for religious conversion, *metanoia*, that is, a fundamental transformation of mind (Pearce, 1971).

Finally there is the task of validating the knowledge (verification) and bringing it to fruition for self and society.

Such processes of discovery may be termed heroic not so much because they parallel the classic stages of separation, initiation, and return of the hero in the monomyth, but because they require inordinate courage in the face of fear. They are truly processes of evolutionary experimentation (see Chapter 3 by Jantsch). They not only entail the possibilities of failure, but also require confronting the truly unknown, as well as the sure knowledge that successful discovery will inevitably upset the established patterns of one's existence. The making of such a discovery implies drastic personal and psychic *change*, in addition to facing the resistance of established authority.

3. EMERGENT CHARACTERISTICS OF A NEW PARADIGM

We have just reviewed the stages that typify the processes of transformation at a number of levels in human systems. The overall pattern that emerges suggests that, generally, when crises are faced and normal modes of problem solving are first seen as inadequate, calls often arise for a "return to tradition" and for achieving desired results by following traditional standards for good performance. Failure here is usually followed by massive "tinkering" with the system. When this does not work, breakthroughs in basic approach are sought, but finding such basic breakthroughs inevitably entails the reexamination and restructuring of the fundamental assumptions and mental maps that underlie conventional wisdom. It leads to transformed perceptions of reality and to altered modes of behavior, and inevitably it stimulates a great deal of resistance as applications of the new perceptions are attempted.

A variety of indicators suggest that our present culture may now be well into such a process of transformation (Markley et al., 1974, pp.

ISBN 0-201-03438-7/0-201-03439-5pbk.

16, 78)—in large part because of increasingly intense human systems problems that lie beyond the ability of traditional approaches to resolve. The recent use of terms such as "the world macroproblem" (Harman, 1970) "the predicament of mankind" (Ozbekhan, 1976), "the crisis of crises" (Platt, 1969a), and "eco-spasm" (Toffler, 1975) calls to mind a planetary situation that was not anticipated in traditional human systems and images.

Characteristics of emerging paradigms can be roughly predicted in at least two ways: (1) on the basis of anomalies in and/or deficiencies of the present paradigm needing to be rectified (e.g., too much emphasis on reductionism and specialization, suggesting a needed complementary emphasis on holistic knowledge processes and on integration of knowledge); and (2) on the basis of emerging values, premises, styles of life, and scientific paradigms (e.g., a surge of interest in Eastern religious and philosophical perspectives coupled with both scientific and lay interest in meditation and other subjective disciplines, suggesting a synthesis of objective and subjective modes of inquiry).

Deficiencies Needing to be Rectified

A number of premises taken for granted during the industrial era now need to be either replaced or complemented (Markley et al., 1974, pp. 63ff.). Three that appear particularly obsolete are as follows:

1. Human progress is synonymous with economic growth and increasing consumption—a notion now challenged by shortages of various key resources and increased pollution.
2. Mankind is conceptually separated from nature and it is the human destiny to conquer and exploit nature—an attitude at distinct variance with modern understandings of ecology.
3. Economic efficiency, specialization, and scientific reductionism are the most trustworthy approaches to fulfillment of human goals—concepts that have raised our standards of living but are dehumanizing our way of life.

If our society tries to endure the coming decades with such industrial-era images and premises—thereby seeking to continue what has been termed the "basic long term multifold trend of Western culture" (Kahn and Wiener, 1967; Kahn and Bruce-Briggs, 1972)—effective governance of human systems may be within reach only at the price of what Bertram Gross (1970) has termed "friendly fascism"—a managed society ruled by "a faceless and widely dispersed complex of warfare-welfare-industrial-communications-police bureaucracies with a technocratic ideology." Although we have already begun to move into such a

ISBN 0-201-03438-7/0-201-03439-5pbk.

future, some authors (e.g., Heilbroner, 1974) are even more pessimistic about the chances for sustaining a democratic approach to governance of human systems due to the increasing intensity of industrial-era systems problems; such analysts see all-out fascism as almost inevitable unless new types of human systems emerge.

Not so clearly under way, but nevertheless becoming visible, are some images and premises that contrast sharply with certain industrial-era concepts. They form an evolutionary vision in which the goals of growth in personal and collective wisdom could replace those of ever-increasing consumption and exploitation, and their successful emergence would necessarily entail a restructuring, both of the overall paradigm of Western culture and of the subset paradigms of several major sciences.

Some Emerging Scientific Developments

Analysis of possibly emerging paradigms (Markley et al., 1974, pp. 133ff.) suggests that a variety of current scientific developments point toward one or more new paradigms in science. Particularly important are two emerging sciences, one dealing with consciousness, the other with dynamic systems, that could contribute strongly to an evolutionary cultural paradigm having many needed characteristics—especially if these sciences are pursued jointly.

Throughout history, mankind has known of various ways to attain higher levels of human awareness, though most have seemed unreliable or have taken years to master. Consciousness researchers are now learning how to combine new tools (biofeedback training, learning theory) with older techniques (meditation, autosuggestion). Their results suggest that we can unlock more rapidly the vast untapped potentials of the human mind, especially its powers of healing and creativity (Masters and Houston, 1966; Academy of Parapsychology and Medicine, 1971; Green et al., 1973; McQuade and Aikman, 1974; Oyle, 1975.).

Mankind has also known that the physical, biological, and human systems making up our reality are somehow organized and sustained. Recent studies of such systems now tend to indicate that similar patterns and principles govern the processes of ecology, of economics, and of the human body and mind. These findings suggest that higher-order systems tend to evolve from lower-order ones, and that no single mode of governance can ensure system stability through time (Platt, 1969b; Vickers, 1970; Metzner, 1971; Lilly, 1972; Bateson, 1972; Pattee, 1973; Laszlo, 1974; see also Chapter 5 by Prigogine).

ISBN 0-201-03438-7/0-201-03439-5pbk.

Taken together, these two disciplines offer insights into facilitating the processes of discovery and problem solving that lie at the heart of the "cycle of transformation" discussed in the previous section. More significant than the specific research findings, however, is the new emphasis itself, which denies none of the conclusions of science in its contemporary form, but rather expands its boundaries. As though in direct response to the present-day prophets of gloom and doom, it bids to offer both a sound problem-solving approach and a new horizon for growth (Bucke, 1960; Harman, 1967; Dunn, 1971; Elgin, 1974; Mitchell, 1974; Markley et al., 1974; Jantsch, 1975).

Presented below are some provisional characteristics that a new image of humankind and the related characteristics that a new consciousness/systems paradigm of science would need if they were to become dominant and effective.

A New Image of Man (Cultural Paradigm)

Though not yet fully shaped, this new image of man (see Markley et al., 1974, pp. 143ff.) tends to:

1. entail an ecological ethic, emphasizing the total community of life and the oneness of the human race;
2. embrace a self-realization ethic, placing the highest value on development of the individual;
3. convey a holistic sense-of-perspective of life;
4. balance and coordinate satisfactions along many dimensions rather than overemphasize those associated with status and consumption;
5. be experimental and open-ended, rather than ideologically dogmatic.

A New Image of Inquiry (Knowledge Paradigm)

Again, though not yet fully shaped, the important emerging features of the science/art of knowing (especially as applied to systems and consciousness; Markley et al., 1974, pp. 139ff.) tend to:

1. be inclusive rather than exclusive, incorporating, for example, wisdom derived from the myths and rituals of prescientific cultures;
2. be eclectic in methodology and epistemology, including, for example, "extrasensory" modes of knowing;
3. lead to a systematization of subjective experience, probably incorporating the concept of hierarchical levels of consciousness—as applied both to humankind and to other forms of matter/energy/life;

ISBN 0-201-03438-7/0-201-03439-5pbk.

4. foster open, participative inquiry in the sense of, for example, reducing the dichotomy between observer, observed, and context of observation;
5. emphasize the principle of complementarity, especially in such issues as causality and acausality, free will and determinism, and material, spiritual, hierarchical, and mutualistic conceptions of systems.

4. CONCLUDING REMARKS

The creative vision of the great men and of the guiding images of the great cultures of history has played a crucial role in the development of our several societies. Many of these guiding images have an unrealized potential for developing a universal sense of world purpose and unification; they have common elements not usually recognized and may not be as mutually contradictory as is commonly believed. Hence, it may be useful to seek out and make visible those common elements in the guiding images of different societies that point in the direction of commonly sensed world purpose. For example, consider (1) the Marxist–Leninist vision of the two stages for the Russian Revolution (first a state-dominated socialism and then—with a withering-away of the state as a result of increased citizen awareness—a transition to a "true communism" in which the ideal "from each according to his abilities, to each according to his needs" would be realized (Chang, 1961, p. 123); (2) the two-level concept of *swaraj* (freedom, both political and spiritual) with which Sri Aurobindo helped inspire India's national freedom movement (Navajata, 1972, p. 105); and (3) the twin mottoes *E Pluribus Unum* (unity out of diversity) and *Novus Ordo Seculorum* (a new order for the ages), which appear with the unfinished pyramid and the transcendental eye on the Great Seal of the United States. Although the political doctrines that accompany such guiding images often seem to conflict at one level, note that each contains a second level of unfinished business. These second levels are not all that different from (1) each other or (2) more evolutionary images, such as the "noosphere" (Teilhard de Chardin, 1961) or "Supermind" (Aurobindo, 1963)—both of which harken back to the Perennial Philosophy (Huxley, 1945) that is the "highest common denominator" of all major world religions. If better understood and applied, the higher-level commonalities and complementarities intrinsic to these great images might well contribute to a new sense of common world purpose, and the integrated pursuit of research on human systems and consciousness is one way this could be facilitated. Another way would be in the political arena, where a new type of leader is starting to emerge (Schwartz, 1974).

ISBN 0-201-03438-7/0-201-03439-5pbk.

If a fascist evolutionary cul-de-sac is to be avoided, the Western world may soon have to undergo an institutional transformation as profound in its consequences as was the Industrial Revolution, and simultaneously a conceptual revolution as shaking as that created by objective science. The prospect of such an evolutionary transformation is both challenging and frightening, for history gives us little hope of avoiding social disruption during times either of transformation or of unsolved crises.

The next decades thus present a formidable but exciting challenge to those who would practice the arts and sciences of human systems—not only because of the societal problems that lie ahead, but also because of the controversies that are likely to surround proposed solutions to these problems. We cannot avoid the crises that face and that await us, but by understanding the nature of our times we may be better able to deal with them.

REFERENCES

Academy of Parapsychology and Medicine (1971). *Varieties of Healing Experience.* Los Altos, Calif.

Aurobindo (1963). *The Future Evolution of Man,* compiled with a summary and annotated by P. B. Saint-Hilaire. Pondicherry, India: Sri Aurobindo Ashram.

Bateson, G. (1972). *Steps to an Ecology of Mind.* San Francisco: Chandler; New York: Ballantine.

Boisen, A. T. (1936). *The Exploration of the Inner World: A Study of Mental Disorder and Religious Experience.* Willet, Clark. Republished 1971, Philadelphia: Univ. of Pennsylvania Press.

Bucke, R. M. (1960). *Cosmic Consciousness,* rev. ed. New York: Dutton.

Campbell, J. (1956). *Hero with a Thousand Faces.* New York: Meridian.

Chang, S. (1961). "The Stateless-Communistic Society of Marxism," in *The Fate of Man* (C. Brinton, ed.). New York: Braziller.

Dunn, E. S., Jr. (1971). *Economic and Social Development: A Process of Social Learning.* Baltimore and London: Johns Hopkins Press.

Elgin, D. (1974). *The Evolution of Consciousness and the Transformation of Society.* Report. Menlo Park, Calif.: Stanford Research Institute.

Erikson, E. (1958). *Young Man Luther.* New York: W. W. Norton.

Fingarette, H. (1963). *The Self in Transformation.* New York: Basic Books.

Goertzel, V., and Goertzel, M. G. (1962). *Cradles of Eminence.* Boston: Little, Brown.

ISBN 0-201-03438-7/0-201-03439-5pbk.

Green, A., Green, E., and Walters, D. (1973). *Brainwave Training, Imagery, Creativity, and Integrative Experiences.* Report. Topeka, Kansas: The Menninger Foundation.

Gross, B. (1970). "Friendly Fascism," *Social Policy*, Nov./Dec. issue.

Harman, W. (1967). "Old Wine in New Wineskins," in *Challenges of Humanistic Psychology* (J. Bugental, ed.). New York: McGraw-Hill.

Harman, W. (1970). *Alternative Futures and Educational Policy.* Report EPRC-6747-6. Menlo Park, Calif.: Stanford Research Institute.

Heilbroner, R. L. (1974). *An Inquiry into the Human Prospect.* New York: W. W. Norton.

Huxley, A. (1945). *The Perennial Philosophy.* New York: Harper.

Jantsch, E. (1975). *Design for Evolution: Self-Organization and Planning in the Life of Human Systems.* New York: Braziller.

Kahn, H., and Bruce-Briggs, B. (1972). *Things to Come.* New York: Macmillan.

Kahn, H., and Wiener, A. (1967). *The Year 2000: A Framework for Speculation on the Next Thirty-Three Years.* New York: Macmillan.

Kantor, R. E., and Herron, W. G. (1966). *Reactive and Process Schizophrenia.* Palo Alto, Calif.: Science and Behavior Books.

Kuhn, T. S. (1970). *The Structure of Scientific Revolutions*, 2nd ed. Chicago: Univ. of Chicago Press.

Laszlo, E. (1974). *A Strategy for the Future.* New York: Braziller.

Lilly, J. (1972). *Programming and Meta-Programming in the Human Biocomputer.* New York: Julian Press.

Markley, O. W., Campbell, J., Elgin, D., Harman, W., Hastings, A., Matson, F., O'Regan, B., and Schneider, L. (1974). *Changing Images of Man.* Report CSSP-RR-4. Menlo Park, Calif.: Stanford Research Institute.

Martin, P. W. (1955). *Experiment in Depth.* London: Routledge and Kegan-Paul.

Masters, R., and Houston, J. (1966). *Varieties of Psychedelic Experience.* New York: Holt.

McQuade, W., and Aikman, A. (1974). *Stress.* New York: Dutton.

Metzner, R. (1971). *Maps of Consciousness.* New York: Collier.

Mitchell, E., (ed.) (1974). *Psychic Exploration: A Challenge for Science.* New York: Putnam.

Navajata (1972). *Sri Aurobindo.* New Delhi: National Book Trust.

Oyle, I. (1975). *Healing Mind.* Millbrae, Calif.: Celestial Art.

Ozbekhan, H. (1976). "The Predicament of Mankind," in *World Modelling: A Dialogue* (C. W. Churchman and R. O. Mason, eds.). North-Holland/TIMS Studies in the Management Sciences, Vol. 2. Amsterdam and Oxford: North-Holland; New York: American Elsevier.

Pattee, H. (ed.) (1973). *Hierarchy Theory: The Challenge of Complex Systems.* New York: Braziller.

Pearce, J. C. (1971). *The Crack in the Cosmic Egg.* New York: Julian Press.

Platt, J. (1969a). "What We Must Do," *Science*, **166**, 1155.

ISBN 0-201-03438-7/0-201-03439-5pbk.

Platt, J. (1969b). "Hierarchical Restructuring," *General Systems Yearbook.*

Polak, F. (1973). *The Image of the Future* (E. Boulding, transl.). Abridged English ed., San Francisco: Jossey-Bass.

Schwartz, P. (1974). "Youth and Tangible Vision," *Christian Science Monitor,* 1 July 1974.

Sullivan, H. S. (1953). In *The Interpersonal Theory of Psychiatry* (H. Perry and M. Gawel, eds.). New York: W. W. Norton.

Teilhard de Chardin, P. (1961). *The Phenomenon of Man* (B. Wall, transl.). New York: Harper.

Toffler, A. (1975). *The Eco-Spasm Report.* New York: Bantam Books.

Vickers, G. (1970). *Freedom in a Rocking Boat: Changing Values in an Unstable Society.* London: Penguin Press; New York: Pelican Books, 1972.

Wallace, A. F. C. (1956). "Revitalization Movements," *Amer. Anthropologist,* **1956**, 264–281.

Wallace, G. (1926). *The Art of Thought.* New York: Harcourt.

ISBN 0-201-03438-7/0-201-03439-5pbk.

12

Evolving Images of Man: Dynamic Guidance for the Mankind Process

Erich Jantsch

The future enters into us in order to transform itself in us long before it happens.

Rainer Maria Rilke

1. MATCHING MICROCOSMOS AND MACROCOSMOS

Superconscious learning, linking the cultural and mankind levels (see Chapter 3 by Jantsch), produces guiding images which man holds of the ultimate relationship between himself and his world—of the relationship between microcosmos and macrocosmos. In process terms, these images express the search for an identity of principles at work in the microcosmos and macrocosmos. This identity has already been recognized by Hermetic philosophy, the oldest comprehensive knowledge system which has come to our time, in its Law of Correspondence: As above, so below; as below, so above. It has remained a recurrent theme through the millennia in many mystical and philosophical schemes. The Vedic concept of the correspondence between *atman*, the true essence of self (or reality within) and *brahman*, the true essence of reality without, is perhaps the most cogent expression of such an identity, though ultimately in terms of structure. But what counts in a process view is the perennial search for the self-realization of this ultimate structure or, in a complementary view, its unfolding.

Living, in a human sense, may then be called the dynamic interpenetration of *brahman* and *atman*, of objective order unfolding in the human world and evolving subjective values to recognize this order by virtue of a basic identity. The encounter of reality challenges the basic self-image of man, those aspects of *atman* which are capable of giving meaning to reality, of bringing out *brahman*. Cognition becomes possi-

Erich Jantsch and Conrad H. Waddington (eds.), *Evolution and Consciousness: Human Systems in Transition.*

ISBN 0-201-03438-7/0-201-03439-5pbk.

ble because, in its innermost core, it turns out to be re-cognition. Understanding reality is not a passive process of adaptation to an absolute which is to be found outside ourselves, but a feedback interaction between a search without and a search within, an "outer" and an "inner" way. Thus the process of searching and activating self-images of man is the real *re-ligio* (see also Chapter 2 by Pankow), the linking backward to our own origins, in which *brahman* and *atman* become one:

> **How do I know the ways of all things at the Beginning?**
> **By what is within me.**
>
> Lao-Tzu, *Tao Teh Ching*

Re-ligio can be experienced in the oneness to which all evolutionary processes link back—be it called the All of Hermetic philosophy, or *shunyata* (the Void, the source of all refined qualities) in Buddhism, or "extensive continuum" (A. N. Whitehead), or *pleroma*, the "nothingness which has all qualities" (C. G. Jung).

The key to this oneness is the mystical experience, for example, in the Christian *unio mystica* or in the Buddhist notion of "going beyond" the human world. But if this order can be experienced directly in a nondualistic way, it can be explicitly expressed only in terms of a complementary or dualistic unfolding of energy. Viewed in this perspective, images of man form an evolving system and become part of the evolution of human consciousness.

2. IMAGE-FORMING THROUGH *RE-LIGIO*

Modern man lives in threefold space: physical, social, and spiritual (cultural). The dualistic unfolding of energy extends over all three aspects of human space. It is characterized by bringing into play the three basic modes of perception or inquiry which I call rational, mythological, and evolutionary (Jantsch, 1975). They may be described in terms of the subject–object or observer–observed relationship: In rational inquiry, subject and object are separated; the observer does not interfere with the observed. In mythological inquiry, subject and object are linked through feedback loops and affect each other; in "mythic dissociation" (J. Campbell) man enters into a relationship with God. In evolutionary inquiry, finally, subject and object become one, both constituting aspects of an unfolding wholeness—*brahmatman*, "Thou art that." At the evolutionary level, we are closest to experiencing an objective oneness;

ISBN 0-201-03438-7/0-201-03439-5pbk.

at the mythological level we interact subjectively with our world; and at the rational level we attempt to objectivate it by separating ourselves from it.

Man's starting platform for building the overall systems of relations with his world is the mythological level of subjectively experienced feedback relations. There, *fear* is a dominant motivating factor—the fear of spirits in an animistic world, the fear of the consequences of violating social laws and taboos, the fear of falling from divine grace. At the evolutionary level, *hope* dominates—the experience of being embedded and conserved in the unfolding of a metasystem, a macrocosmos, and of leading a meaningful life in this superior context. At the rational level, *certainty* rules—but an artificial and often deceptive certainty, obtained at the expense of cutting the umbilical cord linking us to our own world. In our time, where the failure of rational certainty has brought back fear, only renewed hope is capable of giving gracefulness and balance to the lives of humans and human systems.

I view the evolution of mankind images primarily as the continuous attempt to establish *re-ligio* in physical, social, and spiritual space—to link backward to the three aspects of the beginning from which human life unfolds and which the ancient Chinese called the Tao of the Earth, the Tao of Man, and the Tao of Heaven. In Western terms, their reflections may be recognized in the three primal values: the beautiful, the good, and the true (the sacred). When man entered his psychosocial evolution by developing a self-reflective consciousness, the relations with his physical space were most intense, only dimly and subconsciously guided by relations of the spiritual kind. Later in his evolution, man's total space became more richly structured, encompassing all three aspects—physical, social, and spiritual—in increasingly conscious ways.

With *re-ligio* focusing on physical space, the Earth stood for the microcosmos which matched the macrocosmos of the universe. With the emphasis shifting to social space, humanity became the microcosmos; and with *re-ligio* in spiritual space, the microcosmos is contained in self becoming "God within man" and expressing the universal divine principle of the "greater Self." Accordingly, the three aspects of the feedback process of *re-ligio* may also be called *grounding*, *socialization*, and *individuation*. They run in the two directions of integration and differentiation,[1] orchestrating an emerging process hierarchy. In the direction of differentiation, they fall together with ecological,

[1] It should be noted here that in the traditional Western one-way logic, socialization and individuation are normally understood as processes of differentiation only (C. G. Jung). A one-way logic also seems to have limited Hegel's recognition of the same processes. According to a recent, unorthodox interpretation (Habermas,

ISBN 0-201-03438-7/0-201-03439-5pbk.

social, and cultural organization—or history in general (see Chapter 3 by Jantsch). But it is only through the full *re-ligio*, the interpenetration of integration and differentiation, that human life becomes fully creative.

In Table 12.1, I attempt to trace this development as it unfolded from an emphasis on physical space toward an emphasis on spiritual space; certainly, the notions pertaining to physical space changed significantly as social and spiritual space moved into primary focus.

TABLE 12.1. Images of Man Forming through the Processes of Re-ligio in the Three Aspects of Man's Total Space

	GROUNDING (Physical Space)	*SOCIALIZATION (Social Space)*	*INDIVIDUATION (Spiritual Space)*
EVOLUTIONARY LEVEL	Tao of the Earth Vitalism	Tao of Man Humanity Legitimacy *Li*, ethics of whole systems	Tao of Heaven Shunyata God Within Man-becoming-in-universe-becoming Self-realization Alchemy
MYTHOLOGICAL LEVEL	Gestalt Animism Magic	Polytheistic religions Morality Individual ethics	Monotheistic religions Virtue
RATIONAL LEVEL	Regularities Natural laws Physical technology	Behavioral laws Ceremony Social technology	Humanistic psychology Ritual Human technology

Grounding: The *Re-ligio* of Physical Space

In physical space, through the process of grounding, man first brought order into chaos by feeling the gestalt, the life qualities, of objects and

1974), Hegel's conception of history assumes a threefold separation of man from outer nature, society, and his inner nature. Hegel attempted to constitute identity between nature and history, with the latter, as *Geist* (Spirit), continually causing the very separations which it strives to overcome.

ISBN 0-201-03438-7/0-201-03439-5pbk.

forces of nature.[2] They all appeared to him as personified spirits. Such a mythology of natural forces forms the core of *animism.* In order to defend himself and to survive in the chaos of spirit entities, man resorted to magic through "sympathetic action" by placing his feelings into the physical entities themselves—by acting like an animal, the animal spirit could be mastered. Chaldean astronomers tried to identify with stars by exposing their bodies to the rays of a particular star in order to "feel its essence." Sun and moon appeared as "great spirits."

The world became a home for man when he became capable of feeling himself in a nondualistic way (at the evolutionary level) as a manifestation of an all-pervading life-force. The basis for this *vitalism* was an understanding of the periodicity of life processes, including the phases of the moon and the movements of the sun and the planets. Such an understanding is incorporated in Hermetic philosophy as well as in the ancient Chinese philosophy from which emerged the *I Ching*, the *Book of Changes.*

The objectivation of physical space at the rational level came much later with the static Greek cosmos of Parmenides, which gave rise to natural science in a dualistic mood. Regularities in nature became interpreted as natural laws, independent of man, and measurable quantities established the relations between them. For the first time, man saw himself outside nature—he accepted himself as a builder of culture and projected the constructs of his own reason onto his world.

Many spiritual systems of early and more recent cultures have revived the grounding process, the *re-ligio* of physical space, in a new light. Hinduism, in particular, which grew out of a vitalistic basis (Shiva's dance penetrating the earth), places a high value on the human body as the pure vehicle of instinct as well as of the spirit. The arousing of the spirit was linked to the arousing of the Kundalini energy (also expressed as the awakening of the Kundalini snake) and its rising along the spine through the seven chakras which, again, are seen as energy centers of the body as well as of man's social and spiritual life. Yoga, the way to spiritual integration and enlightenment, incorporates much physical exercise toward psychic integration. So do the approaches of the Sufi; the philosophical/mystical system of relations in ancient Chinese internal medicine (especially acupuncture); and certain Japanese ways toward physical-psychic integration and balancing, such as the martial arts (e.g., archery) and Hara. In modern Western culture, bioenergetics—focusing on the liberation and balance of physical, emo-

[2] These beginnings are usually identified with cultures in Southwest Asia around 10,000 B.C. But they were probably preceded by much earlier cultures in Southeast Asia (Sauer, 1952).

ISBN 0-201-03438-7/0-201-03439-5pbk.

tional, and psychic energy flows through work on the body (in contrast to psychoanalysis, which searches only the mind for causes of blocks and disturbances)—has pioneered a new wave of techniques approaching man as a unity of body, soul, and mind. "Listening to the body" is again becoming recognized as a valid source of superconscious learning.

Socialization: The *Re-ligio* of Social Space

With the focus on social space and the process of socialization, man tried to extend his web of personal relations to the macrocosmos through polytheistic religions which were characteristic of earlier cultures. The result, at the mythological level, was *morality*, an *individual ethics* which made man personally responsible to the gods. On earth, social systems were built to match these hierarchical relationships and special links were provided between the divine and the social systems of responsibility by the concept of *legitimacy* of political organization in the name of the gods (e.g., in the Sumerian-Babylonian culture around 3500 B.C., where there also developed the first explicit social class structure of which we know). Sometimes even a direct bond was established by declaring the secular rulers to be direct descendants of the gods (e.g., the Japanese emperors).

An impersonal principle linking the microcosmos of the state to the macrocosmos was introduced in China as *li*, the universal moral order—"Guide the people by virtue and control or regulate them by *li*, and the people will have a sense of honor and respect" (Confucius). The same notion appears not much later in Plato's *Timaios:* "There is only one way in which one being can serve another, and this is by giving him his proper nourishment and motion; and the motions that are akin to the divine principle within us are the thoughts and revolutions of the universe." In our time, the *ethics of whole systems* (see, in particular, Churchman, 1968) attempts somewhat more timidly the same integration of social space toward human systems and ultimately one humanity.

The dualistic objectivation of social space at the rational level of inquiry has found its theoretical framework only in our time in the form of social science with a behavioral credo which underlies *social technology* and its vastly increasing applications. But its way had been paved for a long time by the imposition—through social laws and taboos—of selected behavioral patterns, or *ceremony*, upon social systems and relations.

ISBN 0-201-03438-7/0-201-03439-5pbk.

Individuation: The *Re-ligio* of Spiritual Space

Spiritual space and the process of individuation moved into sharp focus about twenty-five centuries ago with a new phase in the self-reflective development of human consciousness, namely, the emancipation of the human mind through abstract thinking. It contributed greatly to bringing order into physical and social chaos, but it opened up a new chaos of a psychological kind. Dualistic detachment turned wisdom into intellect, and sometimes into sophistry.

In the individuation process, man started out to find himself and to link his own individuality to the divine principle at work in the macrocosmos, to become "man made in the image of God." Thus, God was in turn conceived in the image of man and the corresponding mythology was that of *monotheistic religions.* Their central notion is *virtue*, life in the light of a father-god. The urge toward integration, toward the unification of a world fallen apart in dualistic concepts of good and evil, appears in the promises of redemption in Zoroastrianism as well as in the Judaeo-Christian faith.

But feeling the "God within" at the evolutionary level gives a meaning which needs no further construct to be legitimized. The Gnostics, in their Gospel According to Thomas, tried to reconcile such a feeling with a monotheistic view: "But the Kingdom of the Father is spread upon the earth and men do not see it."

The Gnostic fallacy may be seen in the attempt to bring structural thinking to the evolutionary level. In contrast, the great nontheistic religions, Buddhism and Taoism, express a pure process view which links a dynamic concept of man to the macrocosmos by virtue of man's participation in universal evolution. Buddhism even became a most elaborate evolving system in itself which permits a study of the various stages of spiritual integration, or individuation, within one and the same religious paradigm: Where *Hinayana Buddhism* focuses on the mythological level, seeking precision in seeing things in a dualistic perspective, *Mahayana Buddhism* is geared to *prajna*, or transcendental knowledge, to be found in the *re-ligio* to *shunyata* in which man establishes his oneness with the greater Self; but the highest stage is *Tantra*, the dancing out of energy, the continuous linking backward to the origin in order to energize life, the fluctuation between oneness and dynamic duality—individual life unfolding itself meaningfully as an agent of universal evolution.

Thousands of years ago, *Hermetic philosophy* already anticipated for man the goal of his identification with the greater Self. Medieval *alchemy*, one of its offspring, took the same approach of a "psychological vitalism" with the hope of also mastering physical transformations

ISBN 0-201-03438-7/0-201-03439-5pbk.

in the same spirit, of building a world to the measure of man by subtle regulation of cosmic forces—not an *a priori* impossibility if we think of the small amounts of energy needed for controlling large-scale energy flows in cybernetic systems in which the principles of the microsystem match those of the macrosystem in precise ways. The Cabala, in contrast, took the approach of a "psychological animism" by attempting to master "astral" forces through magic and thereby to achieve physical transformations. The successor to these attempts, modern *technology*, developed at the rational level, first with the same aim of mastering physical nature and later extending to social space. Like the magician in the Cabala, we now have to feed it with our own soul.

The rational approach to individuation—the manipulation of self through *human technology*, or *ritual*—has so far remained peripheral. Psychotherapy in a variety of versions and the techniques developed in the name of humanistic psychology and practiced in "growth centers" and esoteric schools may be regarded as forerunners to future attempts toward the "planning of self" in analogy to physical and social planning. We have not yet reached a stage where the rational pursuit of spiritual values plays a dominant role.

The processes of grounding, socialization, and individuation, as well as the triad of levels of inquiry, may be considered as stratified systems in which meaning is enhanced as one ascends and there is better resolution of detail as one descends. Reductionism in the human world, with the more holistic and evolutionary images of man gradually being cut out and no longer serving as guides to the lower levels of the hierarchy, has been most graphically described in Lao-Tzu's *Tao Teh Ching:*

> Failing Tao, man resorts to Virtue.
> Failing Virtue, man resorts to humanity.
> Failing humanity, man resorts to morality.
> Failing morality, man resorts to ceremony.
> Now, ceremony is the merest husk of faith and loyalty;
> It is the beginning of all confusion and disorder.

ISBN 0-201-03438-7/0-201-03439-5pbk.

3. HISTORY AND EVOLVING IMAGES OF MAN

History may be viewed as the unfolding of *brahman* in the human world. In Chapter 3, I have tried to describe this complex order of processes in terms of "waves of organization" (ecological, social, and cultural), succeeding each other and bringing into play a hierarchy of human systems, orchestrating human life ever more fully. Images of man, as unfolding aspects of *atman*, may then be basically understood as manifestations of internal, coordinative factors at the mankind level which are to be acted out through history. They are archetypes in the ultimate Jungian sense of "psychophysical formal structures (which) are in the final analysis a world-forming principle, a common transcendental ordering factor of existence" (Franz, 1974, p. 203).

Images of man seem to surface primarily at the subjective mythological level where man interacts with the outer world by projecting the order of his inner world onto it and entering into a web of feedback relations, a dialogue, with his environment. Langer (1967, 1972) makes the point that language developed around emotions, not objects. The process of matching order without and order within can only start in such a feedback way, with both sides still malleable in their expressions. But in trying to express *atman* as comprehensively as possible, the guiding images subsequently tend to move toward expression at the nondualistic evolutionary level.

Thus, if we attempt to plot the development of images, as discussed earlier, over time, a curious pattern emerges (see Fig. 12.1): The images enter at the mythological level, obviously as a result of superconscious learning and *in anticipation* of a wave of organization which has not yet reached the corresponding stage of unfolding. From there, still in anticipation, they tend to develop in the direction of integration toward the evolutionary level, thereby preordering a reality which is yet to unfold in the opposite, differentiating direction.[3] This perhaps constitutes the best guarantee of the emerging organization's really becoming a human world: the search for *atman*, the true essence of humanness, interpenetrates with the unfolding of *brahman* in the shaping of the initial phases of each wave of organization. Since everybody shares in *atman*, images of man can become the seed for future cultural patterns. "Knowing that which is to come depends on the backward movement," is also one of the principles on which the *I Ching* is based (*I Ching*, 1967, p. 265).

The focus eventually moves down to the rational level at a later stage, but as a result of conscious learning's taking over, apparently no

[3] In this context, it is interesting to note that the evolution of music, the most human of all arts, has been explained in similar terms by Ernst Bloch (1918): Music successively brought into play three levels, weaving on them ever denser "carpets" approaching an image of an "inner light of humanness" *("Innerlichkeit")* which illuminated and guided the process from the far end. The first and earliest carpet started with "endless singing" and dance and matured to chamber music. The second carpet started with closed song form and matured toward oratorios and Mozartian opera. The third carpet, finally, is of dramatic character and developed from open song form to "transcendental" opera (Wagner), big choral works, and the symphonies of Beethoven and Bruckner. Each new style tends to reinterpret the old patterns of previous carpets in more flexible forms.

ISBN 0-201-03438-7/0-201-03439-5pbk.

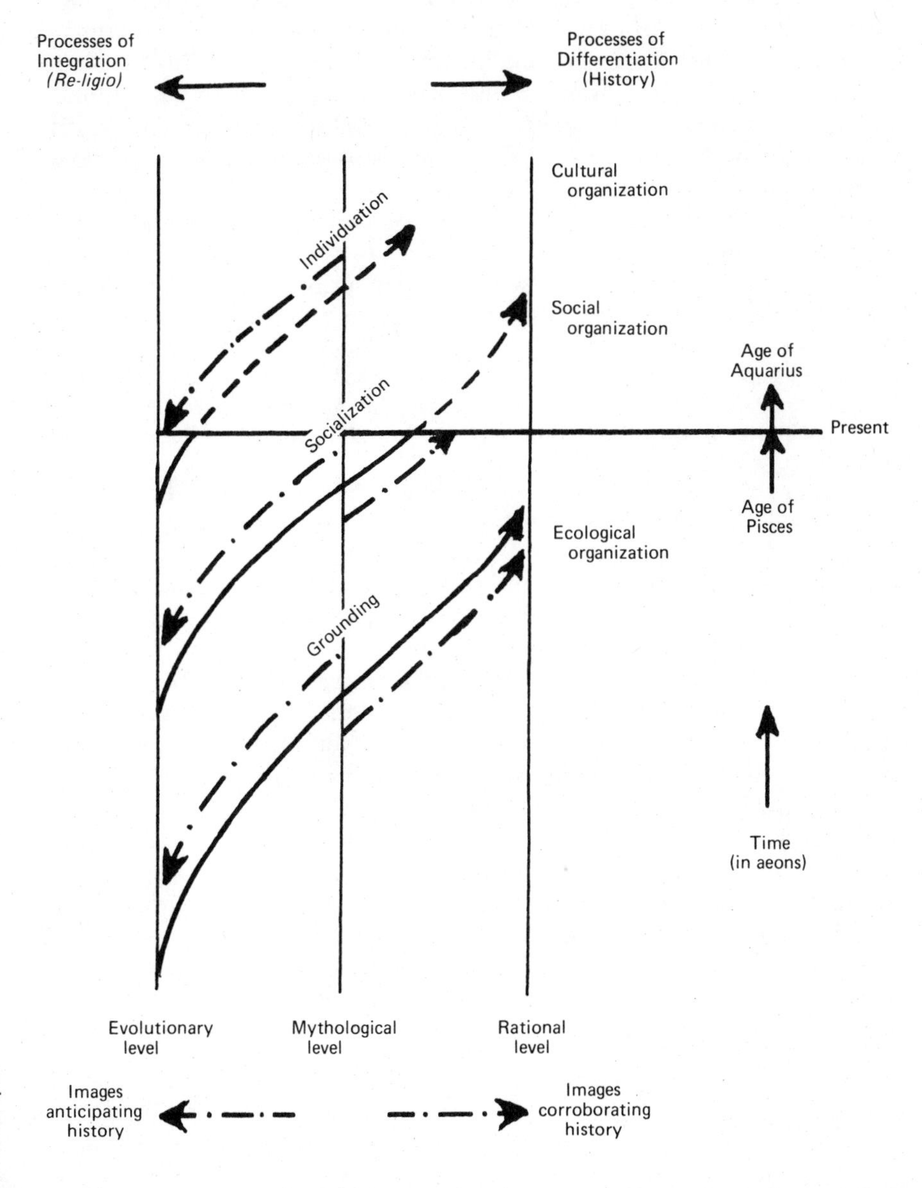

ISBN 0-201-03438-7/0-201-03439-5pbk.

FIGURE 12.1. The interpenetration of processes of differentiation (history) and integration *(re-ligio)* in the human world. Images of man tend to anticipate history through processes of integration, moving from the mythological to the evolutionary level, and to accompany (interpret) history in its moves from the mythological to the rational level.

longer in anticipation of, but synchronized with, the unfolding of the corresponding wave of organization. Rationality in the physical domain appeared with Greek science long after the animistic and vitalistic stages of archaic cultures, and even long after new images of man had illuminated social and spiritual aspects of man's total space and social organization was well under way. The values that go with social planning and behavioral science are about to make their full impact only in our time, concurrent with the movement of social organization into a phase of conscious design.

Thus, images of man, after surfacing primarily as subjective expressions, interpenetrate with history by moving in two directions:

1. through superconscious learning in the direction of integration, providing "*objective normative forecasts*" which anticipate and prefigure the emerging order of the human world and act as powerful self-fulfilling prophecies, thereby guiding history; and
2. through conscious learning in the direction of differentiation as a correlate to history and interpreting the unfolding order of the human world.

Therefore, the physical, social and spiritual aspects of the total image are usually at variance with each other, introducing nonequilibrium (and stimulation toward further evolution).

The emerging hierarchies of order interpenetrate each other in a peculiar way: Whereas the historical process gradually brings into focus human systems of widening scope—from individuals through social systems and cultures to mankind-at-large, thus correlating man with environments which bring in ever more comprehensive aspects of the macrocosmos—the evolution of images of man lets the corresponding microcosmos shrink from the whole earth (grounding) through humanity (socialization) to self (individuation). But this is only a further aspect of the continuous process of matching *brahman* with *atman*, and perhaps also of the evolutionary trend toward greater flexibility of human systems "phenotypes" (see Chapter 3 by Jantsch).

4. CONCLUSIONS

The evolution of self-held images of man through superconscious learning provides a kind of objective, dynamic guidance for the mankind process which reaches far into the future—thousands of years, or aeons. It forms the core of evolutionary experimentation linking the cultural to the mankind level and preparing basic transformations in the noosphere. In contrast, images of man developed through conscious learn-

ISBN 0-201-03438-7/0-201-03439-5pbk.

ing tend to stabilize the existing structure of the noosphere. Thus, particularly in times of transition, an emergent superconscious image may be expected to be in conflict with the predominant conscious image (see Chapter 11 by Markley). It acts as a powerful fluctuation which forces the noosphere toward a new regime, a new basic paradigm of humanness.

Superconscious images span aeons. This means that the image which will guide us through the imminent noetic regime is already with us. The process of individuation, which has been illuminated so brightly at the beginning of the Age of Pisces, will become the great task to be lived out in the Age of Aquarius on whose doorstep we now find ourselves. This insight forms the core of the message that Jung (1961) formulated at the end of his life:

> I think that this is what can be said at the end of our aeon of the Fishes, and perhaps must be said in view of the coming aeon of Aquarius (the Water Bearer), who has a human figure and is next to the sign of the Fishes. This is a *coniunctio oppositorum* composed of two fishes in reverse. The Water Bearer seems to represent the self. With a sovereign gesture he pours the contents of his jug into the mouth of *Pisces austrinus*, which symbolizes a son, a still unconscious content. Out of this unconscious content will emerge, after the passage of another aeon of more than two thousand years, a future whose features are indicated by the symbol of Capricorn: an *aigokeros*, the monstrosity of the Goat-Fish,[4] symbolizing the mountains and the depths of the sea, a polarity made up of two undifferentiated animal elements which have grown together. This strange being could easily be the primordial image of a Creator-god confronting "man," the Anthropos. On this question there is a silence within me, as there is in the empirical data at my disposal—the products of the unconscious of other people with which I am acquainted, or historical documents. If insight does not come by itself, speculation is pointless.

REFERENCES

Bloch, Ernst (1918). *Zur Philosophie der Musik.* Republished 1974, Frankfurt-am-Main: Suhrkamp.

Churchman, C. West (1968). *Challenge to Reason.* New York: McGraw-Hill.

Franz, Marie-Louise von (1974). *Number and Time: Reflections Leading toward a*

[4] In the original text, the following footnote appears: "The constellation of Capricorn was originally called the 'Goat-Fish'."

ISBN 0-201-03438-7/0-201-03439-5pbk.

Unification of Depth Psychology and Physics (A. Dykes, transl.). Evanston, Ill.: Northwestern Univ. Press.

Habermas, Jürgen (1974). "Können komplexe Gesellschaften eine vernünftige Identität ausbilden?," in *Zwei Reden.* Frankfurt-am-Main: Suhrkamp.

I Ching, The, or *Book of Changes.* (1967). Richard Wilhelm, transl.; rendered into English by C. F. Baynes; 3d ed. Princeton, N.J.: Princeton Univ. Press.

Jantsch, Erich (1975). *Design for Evolution: Self-Organization and Planning in the Life of Human Systems.* New York: Braziller.

Jung, Carl Gustav (1961). *Memories, Dreams, Reflections.* New York: Vintage.

Langer, Suzanne K. (1967, 1972). *Mind, an Essay on Human Feeling,* 2 vols. Baltimore and London: Johns Hopkins Press.

Sauer, Carl O. (1952). *Agricultural Origins and Dispersals.* New York: American Geographical Society.

ISBN 0-201-03438-7/0-201-03439-5pbk.

Concluding Remarks

Conrad H. Waddington

By the time he invited me to join him as co-editor of this volume, Erich Jantsch had already got a pretty good idea of what he wanted to include in it, and had in fact already spoken to a number of possible contributors. We discussed the general plans fairly thoroughly, but the work of selecting the actual contributors, and persuading them to write something, was undertaken by Erich. He went ahead with what, I am sure, was exceptional rapidity, and the whole volume was assembled in typescript before I had got rid of some other commitments sufficiently to give much attention to it. I realize now that the word "Jantsch," besides being a surname of an individual, is, or ought to be, the name of a certain quality—something allied to zest, verve, dash, élan, combined with efficiency and accomplishment. It was with this quality of "Jantsch" that the task was completed, of translating this volume from an idea into a set of articles ready to be shipped to the printer.

I feel that the only way I can hold my head up as co-editor is to make some post-hoc comments on the essays which Erich has assembled and put in order.

I will start with those which are closest to the natural sciences, namely, Prigogine's discussion of systems which are far from equilibrium; Holling's account of concepts related to equilibrium in the context of ecology; and Abraham's discussion of macrons, morphogenesis, and catastrophes. I shall then move on to the question whether the kind of systems we are interested in can be adequately discussed in terms of information, and if not, what other terms would be more appropriate to the phenomenon of self-transcendance on which Pankow and others lay so much stress in the first part of the volume. Finally, I shall turn to some remarks about the problems which are taken up in Part III of the volume.

I approach the first group of problems from a professional background which differs from that of other writers. One of the major types of biology with which I have extensive experimental and theoretical acquaintance is the analysis of embryonic development; and embryos provide exceptionally favorable material for studying the problems with

ISBN 0-201-03438-7/0-201-03439-5pbk.

Erich Jantsch and Conrad H. Waddington (eds.), *Evolution and Consciousness: Human Systems in Transition.*

which these three authors are engaged. They are, first, quite obviously not in a state of equilibrium. A fertilized egg insists, one might say, on changing. The only way to stop it changing is to kill it; that is to say, to alter it from being an open system to being a closed one, by preventing the flow of energy, oxygen, and perhaps other raw materials through it. Moreover, in its normal life, as it changes it also very obviously becomes more complex, both in its overall shape and in the number of discernibly different parts which can be discriminated within it. For these reasons I do not think any serious embryologists have considered that the second law of thermodynamics can be applied in any simple way to their subject material, in spite of what classical physicists might say. In my own case, I was fortunate enough to learn my embryology—or rather to teach it to myself, since as a student I had studied geology—at a University and at a time when the most creative young physicists were people like Blackett, Cockroft, Wilson, Dirac, and others, who were themselves so busy revolutionizing classical physics that they were hardly tempted to attempt to impose it as a rigid dogma on biology.

One can very easily examine large numbers of individual embryos belonging to the same species, and they go through their developmental performance quite quickly. This makes them much more advantageous for the study of the behavior of nonequilibrium systems than the much more intractable material which Holling has discussed. Even to get a reasonably complete description of an ecosystem is a matter for a team of workers, rather than a single person; at least a botanist, to identify the plants, an entomologist, a protozoologist, a bacteriologist, a general zoologist, a physiologist, and I do not know how many more. Moreover, the qualities of resilience and stability which Holling finds in the behavior of ecosystems require several years, if not decades, to become manifest. The number of ecosystems where stability and resilience have been adequately studied in even *one* example is pitifully small. The "stability" notions developed in embryological theory are, therefore, more penetrating and complete than any which ecologists have yet had time to reach.

Embryos, like ecosystems, are multifactorial. They have many characteristics and they are affected by many factors, for instance, a multitude of genes and many environmental conditions. Changes in embryos therefore have to be symbolized by trajectories in multidimensional phase space, in a manner similar to that used by Holling to describe his ecosystems. Any orderliness we can discover in an embryonic system can be described in terms of constraints on the courses of these trajectories. These constraints can be visualized as attractor surfaces. If the system starts from any condition, which will be represented by a point in the multidimensional phase space, the trajectory from that point will first of all move to the nearest attractor surface, and then move along

ISBN 0-201-03438-7/0-201-03439-5pbk.

that surface. To deal with embryos, whose changes and stabilities have been much more thoroughly studied than those of ecosystems, we have to describe attractor surfaces of more complex form than the simple basins which Holling spoke of and illustrated in his Figure 4.2.

The first and most obvious type of change in an embryo is that it develops into an adult form; this is relatively long-lasting, but is in fact always undergoing slow processes of change, which lead eventually to senescence and death. This implies that the phase space in which the system is modeled contains a surface with a general slope which will guide any trajectory toward the adult state, and final death. However, we have also to take account of the fact that different parts of an embryo develop into different organs—liver, kidney, brains, muscles, and so forth. This situation can be described by supposing that superimposed on the general slope is a radiating system of valleys, which direct some trajectories to move along toward the kidney, another set to move along toward liver, and so on. An attractor surface modeled in such a manner is called an epigenetic landscape.

A description in these terms suggests many questions which it might be profitable to study. At what point do various valleys branch off from one another? Do two valleys which have once become separate from one another ever later come together and fuse again? But perhaps the most important questions for most practical purposes relate to the shapes of the valleys in cross section and the height of the mountain watersheds between them. There is no reason why a valley should not have the shape of a very narrow canyon with precipitous and possibly high walls; or alternatively be characterized by the gentler contours, leading down to broad water meadows through which the river meanders, which are characteristic of what geologists would regard as an old, mature earth form.

To avoid having to use the only metaphorical name of a "valley," for what is really a characteristic of an attractor surface in a multidimensional space, I have coined the word "chreod." The cross-sectional shape of the chreod describes the reaction of the system to fluctuations affecting it. In a chreod with a canyon-like shape, it will be very difficult to divert the developing system from the very bottom of the valley. If this is done by a strong enough influence, the system will immediately find its way back to the bottom as soon as the influence ceases. Such a system is very stable in Holling's terms. On the other hand, if the chreod has the shape of a broad river floodplain, it will be very easy to divert the system from the very lowest point and it will return there only after meandering at random for quite some distance. This is a system with little stability in Holling's terms.

On the other hand, the resilience of the system depends not on the shape of the river valley or chreod, but on the height of the watersheds

ISBN 0-201-03438-7/0-201-03439-5pbk.

on each side of it. These indicate the maximum fluctuation which the system can absorb while remaining within the same chreod. Anything greater than this will push the system out of this chreod over the watershed. Now Holling suggests that it is very unstable systems (i.e., those corresponding to broad valleys), which are most resilient (i.e., have the highest watersheds). In a very general way, in epigenetic landscapes comparable to old matured earth forms, this may be true; but the connection is not necessary. In fact, in newly formed landscapes, developed for instance in regions of the earth recently subject to considerable uplift, one may find deep valleys which are very narrow in a cross section, such as the Grand Canyon. If one is trying to design and produce in a relatively short time a system in which it is of the first importance that it remain in its own chreod and not be pushed over a watershed into something quite different, then the simplest plan would be to make a very deep valley and a narrow one. The objection to this is not that its stability necessarily robs it of resilience, but that it allows very little variation among the individuals in a population passing along it. One could say that it produces a great deal of turbulence in the stream at the bottom. The social ideal would seem to be to allow a great deal of individual variation, in a maze of meandering streams in a flat valley bottom, but at the same time to have high watersheds on each side to prevent the system's being flipped out of that chreod into some unknown country.

This raises the question what happens outside the watershed of a particular chreodic valley. In biological systems all the alternative combinations of factors have been subjected to long-term natural selection. By now, most of the choices are between well-formed chreods leading to definite end results. However, even in biology one can find cases where this is not so; for instance, when cells become cancerous, or chaotic when explanted into tissue culture. There is no theoretical reason why, when a fluctuation carries a nonequilibrium system over its resilience boundary (or watershed of the chreod), there should be any other attractor surface available.

In my opinion, therefore, the summary of Holling's argument presented by Jantsch in his points 6 and 11 of his summary at the end of the Introduction is really based on a confusion between two different types of stability. When a system is "near the boundaries of a stable regime" (i.e., near the maximum sustainable nonequilibrium), there may indeed be short-term instability in that the system is altered quite considerably by small fluctuations, but in these circumstances there is a danger that even quite a small fluctuation will carry it across the watershed, so that it falls under the influence of another attractor. This small-scale instability near the boundary can be visualized as a relatively

ISBN 0-201-03438-7/0-201-03439-5pbk.

flat top to the watershed ridge, as one often finds in old well-rounded countrysides, rather than the watershed having the form of a sharp ridge or crest. In my opinion, the short-term instability that is desirable is, on the contrary, that characteristic of the floor of a broad valley. Here again the surface is relatively flat and small fluctuations can produce large displacements, but here there is no danger of the system's toppling over from one region into a quite different one. It is in these circumstances that high instability goes with great resilience, the high instability being the result of the flatness of the immediate surroundings and the high resilience the effect of the high walls of the distant watersheds.

The work of Abraham, and that of Hans Jenny, offer the possibility of an experimental empirical approach to some of these questions. For instance, one may study the influence of boundary conditions such as the shape of a metal plate, or dynamic conditions such as systems of vibration, on the production of patterns which are either orderly or in some cases lack any evident trace of order. Abraham offers, as he himself points out, some wild and rather far-reaching speculations about possible applications of this type of thinking. I think these are useful suggestions about the importance which a proper understanding of the arising of order might have for many areas of experience. However, what one needs most at the present time is some good basic theory, some structure of concepts and methods of approach to the problem, comparable say to René Thom's theory of the catastrophe surfaces by which a multidimensional phase space is divided up into separate regions.

I shall now turn to the question of self-transcendence which was introduced by Pankow and discussed in many of the articles. I think a first step toward considering this very "deep" topic is to consider an appropriate way of referring to the specificity which is such a characteristic of biological organisms and all their parts, from major to minor. One species is specifically different from another, and so in a different way is one individual of a given species specifically different from any other individual. An organ such as a liver is specifically different from another organ, such as a kidney. Each type of cell is specifically different from each other type of cell, and so on down to the smallest cellular organelles, and eventually to the genes. In recent years it has been usual to discuss these differences in terms of *information*, the word information being given a technical meaning defined by Shannon and Weaver, in which it refers to specific differences of kind within a defined universe of variations.

Recently it has been becoming widely recognized that such a world of information is essentially static, in that a system cannot engender

ISBN 0-201-03438-7/0-201-03439-5pbk.

new information from within itself; the only type of spontaneous change possible is a loss of information. This is clearly a highly inappropriate type of formulation to use in connection with developing embryos or evolving biological organisms. These certainly and without question exhibit what might be considered the most elementary aspect of self-transcendence, namely, increase in the degree to which they express specificities. Systems which exhibit this property can be described if one expresses the specificities not as statements or information, but rather as *instructions* or algorithms. As computer scientists have so clearly demonstrated, a system containing quite a few instructions, including instructions to repeat an action, can produce results which appear to be of fantastic complexity in terms of information. Some sets of instructions, at least, are deviation amplifying in the sense of Maruyama. These can bring about not only relatively trivial increases in specificity, but can carry a system out of a region of phase space dominated by one attractor surface over a catastrophe surface or watershed into another chreod.

The concept of instructions is much more important for all aspects of biology than that of information. It is, however, not sufficient in itself as Pankow, Jantsch, and others point out. Biological systems, in their genetic and evolutionary processes, transcend themselves in a way comparable to that in which a natural language can discuss its own structure (and becomes in doing so a metalanguage), a possibility which is not open to completely formal language. Pankow speaks of this additional phase of self-transcendence as a form becoming a "gestalt." A similar point was made, particularly by Howard Pattee, during a series of discussions on theoretical biology which I organized a few years ago at the Villa Serbelloni in Italy. There the notion was put most aphoristically by pointing out that we regard certain biological molecules as messages; that is to say, we consider them as conveying instructions of a kind comparable to the instructions which can be given in a natural language. I am sure that this is an extremely important manner of regarding living things and their evolution, but I think it still needs a great deal of further working out, both as to its biological basis and its consequences. I confess that I am not at all clear about the relations between sets of instructions in general, those sets which may lead to deviation-amplifying systems, and those which give rise to the self-transcendence characteristic of natural languages. But we are certain that at least some sets of instructions can achieve this transcendence, so we can perhaps leave it to the professional philosophers to decide what characteristics these sets must necessarily possess.

It is more important, in connection with the topics discussed in Part III of this volume, to consider the consequences of such a mode of

ISBN 0-201-03438-7/0-201-03439-5pbk.

approach to the problems of development and evolution. I would tentatively advance the following arguments. The specificities, we have said, should be expressed in terms of instructions. Instructions are necessarily instructions to behave in certain manners, that is, to alter things in some way or other. Any alteration of a situation must always have a characteristic corresponding to a "value" for *some* system of assessment; for instance, a genetic change of instructions for the synthesis of a particular protein, or for carrying out a particular type of behavior, will have value from the point of view of natural selection. This leads to the conclusion that the expression of specificities in terms of instructions necessarily involves us in normative thinking. Marney and Schmidt point out that a normative-theoretic mode of inquiry demands a characteristic contemporary modification of previous scientific methods, but they do not point out just why it does so. According to the argument above, the demand arises because of the inadequacy of purely informational theories to deal with developing and evolving systems.

Since all biological systems contain a multiplicity of instructions, it seems natural, if not inevitable, that they will be involved in a multiplicity of value systems. The existence of what Marney and Schmidt think of as "simultaneous but antithetical desiderata," and the importance of taking multiple viewpoints insisted on by Maruyama, follow very naturally from this point of view, and I am sure both are very important. Jantsch, in his phrase "symbiotization of heterogeneity," expresses very well the basic principle we have to follow in this connection (Introduction, point 10). As he points out immediately before this, it is only by meeting in a balanced way a number of different, and in some cases opposing, demands, that evolution can bring about an increasing harmony between the fundamental nature or genotype of an organism and its environment. I think he is right also in his contention that the most urgent problem of immediate human evolution is to improve this harmony, particularly in respect to the social and cultural environments. Here we shall need a great deal more of what Marney and Schmidt refer to as "a third thinking for the sake of new modalities of thought. This is thought consciously aimed at optimization of the cognitive modality, and realization of successively more flexible, more general, more satisfactory modes of comprehension and action." That is, a conscious, devoted, but critical attempt to create new guiding images of man and the future of the kind discussed by Markley. I believe that the new ways of thinking which are being developed in terms of instructions and value-inclusive normative modes are beginning to provide methodologies capable of achieving the necessary, challenging, and inspiring tasks.

ISBN 0-201-03438-7/0-201-03439-5pbk.

Author Index

Numbers set in *italics* indicate pages on which complete literature citations are given.

Subject Index

ALSO OF INTEREST

René Thom,

STRUCTURAL STABILITY AND MORPHOGENESIS

"Fowler's translation of Thom reads like a book written originally in English, and a fair sprinkling of the figures in the French edition have been replaced by clearer ones. Thom has taken advantage of the appearance of this translation to revise the material. . . . Thom's book remains immensely stimulating and imaginative; the two years since my last review (of 'Stabilité Structurelle et Morphogénèse') have not changed my opinion that it sets out a major intellectual advance of the century which no reader, even moderately versed in mathematics, can ignore."

— C. W. Kilmister,
Times Higher Education Supplement

1975, (2nd printing with corrections, 1976) xxvi, 348 pp., illus.; ISBN 0-850-39278-5, ISBN 0-805-39279-3 (pbk.)

Harold A. Linstone and Murray Turoff (eds.),

THE DELPHI METHOD: TECHNIQUES AND APPLICATIONS

"A day seldom goes by in which I do not receive a request for information on the Delphi method, and I am delighted that at last a compendium has been put together that fills the clear need indicated by such inquiries. While its principal area of application has remained that of technological forecasting, it has been used in many other contexts in which judgmental information is indispensable. These include normative forecasts; the ascertainment of values and preferences; estimates concerning the quality of life; simulated and real decision-making; and what may be called 'inventive planning'."

— from the Foreword by Olaf Helmer,
University of Southern California

"This book is the first monograph dealing with the theory and application of the Delphi Method. It may well become the classic reference work for all those concerned with technological forecasting or social sciences where no hard data are available. A number of such studies are described in detail; full consideration is also given to computerization of the technique and its inherent limitations."

— Neue Zürcher Zeitung

1975, xx, 620 pp., illus.; ISBN 0-201-04294-0, ISBN 0-201-04293-2 (pbk.)

Erich Jantsch, born in 1929, received his Ph.D. from the University of Vienna in 1951. He has made important contributions to science and philosophy and held visiting appointments, chiefly at the University of California, Berkeley, and also at Massachusetts Institute of Technology, Portland State University, Technische Universität Hannover, Universität Bielefeld, Technical University of Denmark, University of Paris, Institute of Advanced Studies, Vienna, University of Lund, Sweden, and the Graduate School of Economic and Social Sciences, St. Gallen, Switzerland.

The author's previously published books include Technological Forecasting in Perspective, Perspectives of Planning (ed.), Technological Planning and Social Futures, and Design for Evolution. Dr. Jantsch is one of the founding editors of the journal, *Technological Forecasting* and an associate editor of *Management Science*.

Conrad H. Waddington, C.B.E. (1905-1975), a renowned biologist, educator, and author, was born in Evesham, England, and educated at Clifton College, Cambridge University, Trinity College, Dublin, Prague, Geneva, and Aberdeen. A prolific author, he contributed to scientific journals and published Organisers and Genes, Science and Ethics, The Scientific Attitude, Principles of Embryology, The Strategy of the Genes, Biological Organization, The Ethical Animal, The Nature of Life, New Patterns in Genetics and Development, How Animals Develop, Biology for the Modern World, Principles of Development Differentiation, Behind Appearance, Evolution of an Evolutionist, Tools of Thought for Complex Systems, and served as editor of Towards a Theoretical Biology, Vols. 1-4. Professor Waddington was Lecturer in Zoology, Strangeways Research Laboratory, Cambridge University; a Fellow of Christ's College; Buchanan Professor of Animal Genetics, and Director of the School of Man-Made Future, University of Edinburgh; Visiting Einstein Professor, State University of New York, Buffalo; and President, International Union of Biological Sciences. He was a Fellow of the Royal Society and of the New York Academy of Sciences as well as a Fellow at the Center for Advanced Studies, Wesleyan University. Dr. Waddington was also a member of the American Academy of Arts and Sciences.

Evolution and Consciousness: Human Systems in Transition
Here are some of the emergent concepts and formal approaches which seem capable of grasping some of the important characteristics of evolving systems and coherent behavior, or "continuity in change." The contributors to the volume argue that the evolving world of human systems is characterized by the same aspects of imperfection, nonequilibrium and nonpredictability, of differentiation and symbiotic pluralism which seem to govern life in all its manifestations. These new concepts and formal approaches seem to become a powerful integrative force in our understanding of how a totality moves, from physical, biological, and sociobiological to sociocultural aspects of evolution. What is real in this exciting drama of continuous self-renewal and self-realization? Self-bounding processes rather than the elusive and transitory structures which arise from their interaction in ever-changing forms and complexities.

Evolution and Consciousness is one of the first, still rare, truly transdisciplinary books: it deals with a totality, not a sector out of it. Therefore, it defies any disciplinary labeling. It is a scientific book, yet also deals with topics until now reserved for books of mysticism and poetry. It bridges the gap between science and other forms of knowledge. It deals not just with scientific questions, but with existential questions which concern all mankind, such as the meaning of life and the evolutionary significance of human design and action. It challenges the whole dominant Western world view: process thinking instead of structural thinking, dynamic instead of static, evolution instead of permanency.

Addison-Wesley Publishing Company, Inc.
Advanced Book Program
Reading, Massachusetts